SpringerBriefs in Energy

# Computational Modeling of Energy Systems

**Series Editors**

Computer aided engineering has become indispensable in all major engineering sectors to shorten development times, limit expensive testing, and understand technological working principles as well as failure mechanisms.

In science, virtual in silico laboratories play a very similar role and have become an important means to testing research hypotheses and running virtual experiments. Many of the world's societies are currently transforming their energy systems to renewable energy sources, decentralized energy landscapes, and smart grids. The success of this transition is vital for the establishment of an ecologically, economically and socially sustainable future. The details of the future energy mix remain to be established and science must provide options for future energy systems along with the means to assess these options. Computer aided engineering and assessment capabilities for the novel components of those systems and the systems themselves need to be developed and new generations of engineers and scientist trained to use them effectively.

This subseries puts a spotlight on advanced computational and theoretical methods, tools, and frameworks for the design, analysis, optimization, and assessment of a diverse range of energy technologies and systems. The intention is to make the methods transparent and to allow engineers and scientists from different disciplines to enter the field of energy research enabling them to perform meaningful simulations for the advancement of clean and secure energy systems.

Rabaï Bouderhem

# A Computational Framework for Energy Policy Coherence

Springer

Rabaï Bouderhem
Prince Mohammad bin Fahd University
Al Khobar, Saudi Arabia

ISSN 2191-5520 ISSN 2191-5539 (electronic)
SpringerBriefs in Energy
ISBN 978-3-032-22977-9 ISBN 978-3-032-22978-6 (eBook)
https://doi.org/10.1007/978-3-032-22978-6



This Springer imprint is published by the registered company Springer Nature Switzerland AG
The registered company address is: Gewerbestrasse 11, 6330 Cham, Switzerland

*To my family.*

# Preface

This book emerges from a conviction that governance systems must evolve to match the complexity of the problems they address. Energy policy operates within complex adaptive systems (CAS) characterized by institutional fragmentation, dynamic interactions, and emergent properties that defy simple optimization. Yet computational approaches remain resolutely linear. They assume equilibrium conditions, privilege measurable metrics, and treat policy analysis as a technical problem with algorithmic solutions.

Such failure is not technological but conceptual. Machine learning models, natural language processing tools, and optimization algorithms demonstrate impressive technical capabilities. They achieve high accuracy rates, process vast document corpora, and generate actionable insights. Policy incoherence, however, persists. Institutional fragmentation continues. Marginalized populations and Indigenous communities, those most affected by energy policy decisions, remain systematically excluded from computational systems that increasingly shape the modern era of governance.

I argue that overcoming these failures requires a new analytical lens, bridging complexity science and ethical frameworks for computational governance. This synthesis addresses both the systemic dynamics that resist transformation through technical sophistication and the ethical imperatives that expose how current applications perpetuate systematic exclusions. I draw upon well-established scholarly traditions in complexity economics, organizational theory, epistemology, information ethics, responsible artificial intelligence, and energy justice. These frameworks illuminate distinct yet complementary dimensions of computational governance challenges.

The core diagnostic contribution is the governance-data-inequality nexus. I define this nexus as a self-reinforcing mechanism wherein institutional fragmentation generates data infrastructure constraints, enabling algorithmic bias that legitimizes continued fragmentation. Ultimately, this vicious circle obstructs policy coherence. When energy ministries and environmental agencies operate in silos, they create path-dependent data architectures that become progressively entrenched. Computational applications trained on these datasets inherit their epistemological limitations. The unequivocal consequences are a systematic exclusion of informal knowledge, off-grid systems, community-managed resources, and diverse energy practices reflecting varied socioeconomic contexts. Not through technical failure but through perfect execution of objectives that encode existing power structures and knowledge hierarchies.

Foundational scholarship has thoroughly theorized individual dynamics across complexity science and ethical frameworks. However, I believe their integration for analyzing algorithmic policy coherence represents an original contribution and perspective. Here, I operationalize the theoretical synthesis through a comprehensive computational framework designed on ethical principles as primary design constraints. The BRIDGE-E (Bias-Resistant Integration of Data, Governance, and Equity for Energy) framework represents a testable proposition with comprehensive evaluation protocols designed to assess technical performance, institutional adoption, and ethical outcomes across diverse governance contexts.

I also argue that contemporary governance confronts a fundamental challenge. Computational systems increasingly shape policy decisions, resource allocation, and democratic processes, yet governance mechanisms remain inadequate for addressing their systemic effects. Artificial intelligence transitions from research to infrastructure deployment ranging from welfare, energy to finance and healthcare. This means imperatives extend beyond technical performance to encompass accountability, transparency, and democratic legitimacy. Recent regulatory initiatives, including the most advanced ones like the European Union's AI Act, reflect a consensus and a recognition that computational systems deployed in high-stakes settings require robust governance mechanisms. However, regulation alone proves insufficient without architectural innovation enabling compliance with ethical principles while maintaining computational utility. Effective governance must itself become a complex adaptive system. This transformation necessitates more than bureaucratic layering. It requires building institutional capacity and engaging distributed knowledge across diverse communities. We need to support deliberative processes that respect both technical constraints and ethical imperatives. Governing complexity requires complexity.

This approach treats marginalized voices and Indigenous populations not as stakeholder groups requiring consultation but as knowledge holders whose epistemological frameworks must shape computational systems from inception. Policy coherence in complex adaptive systems cannot emerge from optimization over partial knowledge. It requires planning distributed intelligence across diverse knowledge systems—academic, practitioner, traditional, Indigenous—through deliberative processes that operationalize ethical principles. This addresses epistemic exclusions operating at multiple levels: credibility deficits that systematically discount certain forms of knowledge, and conceptual gaps that prevent adequate articulation of marginalized experiences. The proposed framework enables communities to co-shape the conceptual categories and knowledge structures computational systems employ.

The research objectives encompass multiple dimensions. First, I develop a diagnostic framework explaining persistent patterns of policy incoherence that practitioners encounter but conventional analytical approaches struggle to characterize.

Second, I derive architectural requirements through empirical analysis of how complexity dynamics and ethical failures manifest when institutions attempt computational governance innovation. Third, I propose a testable computational framework building on the complexity-ethics principles through specific technical mechanisms designed for scarce resources and institutional fragmentation, characteristics of developing countries. Fourth, I present a holistic validation protocol enabling rigorous assessment across technical, institutional, and ethical dimensions.

This work aims to shift computational governance away from technocratic optimism, and the assumption that sufficient algorithmic sophistication can overcome structural barriers. It embraces the development of complexity-aware and ethically grounded frameworks. This shift does not reject the transformative potential of computational methods; rather, it takes seriously the systemic dynamics that constrain them and the ethical imperatives that must direct their design. If successful, computational applications could fulfill their transformative promise. Their role would shift from optimizing existing systems to framing the institutional learning and epistemic inclusion essential for more just, coherent, and effective governance. This requires rethinking the objectives and the knowledge computational governance itself values. This is how we will achieve legitimate, equitable, and effective outcomes.

Al Khobar, Saudi Arabia Rabaï Bouderhem
January 2026

**Competing Interests** The author has no competing interests to declare that are relevant to the content of this manuscript.

# Contents


**Part I Foundations**

**1 Bridging Computational Techniques and Energy Policy Coherence** . . . . . 3
1.1 Background and Problem Statement . . . . . 4
1.2 Theoretical Framework and Research Gap . . . . . 7
1.3 Research Objectives and Methodological Approach . . . . . 9
1.4 Methodological Framework and Significance . . . . . 10
References . . . . . 13

**2 Complexity Science and Ethical Frameworks for Computational Governance** . . . . . 17
2.1 Policy Coherence: Conceptual Ambiguity and Political Reality . . . 18
2.2 An Integrated Analytical Lens: Complexity, Ethics, and the Computational Challenge . . . . . 19
2.2.1 Complexity Science, Economics, and the Theory of Complex Adaptive Systems (CAS) . . . . . 19
2.2.2 Ethical Foundations: From Epistemic Justice to Responsible AI . . . . . 20
2.3 Epistemological Challenge and Translational Barriers . . . . . 21
2.4 OECD Approach: Policy Coherence for Sustainable Development (PCSD) . . . . . 24
2.4.1 Core Definition and Purpose . . . . . 24
2.4.2 The Eight Institutional Building Blocks . . . . . 24
2.4.3 The 2019 Council Recommendation: A Three-Pillar Structure for Action . . . . . 25
2.4.4 Practical Application: Evidence from Romania . . . . . 25
2.4.5 Implications for Energy Efficiency (SDG 7) . . . . . 26

2.5 Computational Applications in Policy Analysis: Capabilities and Limitations ... 26
2.5.1 Concrete Illustrations for Energy Efficiency (SDG 7) ... 27
2.5.2 Synthesis: Translational Barriers in Current Applications ... 28
2.5.3 The Architectural Imperative ... 28
References ... 29

3 An Adapted Scoping Review Protocol for Complex Fields ... 33
3.1 Methodological Rationale ... 33
3.2 Systematic Scoping Review: Design and Implementation ... 34
3.2.1 Research Questions ... 35
3.2.2 Search Strategy and Database Selection ... 36
3.2.3 Inclusion and Exclusion Criteria ... 38
3.2.4 Selection Protocol and Quality Assessment Procedures ... 38
3.2.5 Synthesis and Thematic Analysis ... 42
3.3 Data Collection Sources ... 42
3.4 Methodological Limitations ... 43
3.5 Ethical Considerations in Research Design ... 45
References ... 45

**Part II Empirical Analysis**

4 Computational Applications in Energy Policy Analysis: Empirical Patterns and Translational Barriers ... 49
4.1 From Application Catalog to Barrier Diagnosis ... 49
4.2 Empirical Patterns: Three Categories of Computational Applications ... 50
4.2.1 Natural Language Processing: Semantic Analysis of Policy Discourse ... 50
4.2.2 Agent-Based Modeling: Behavioral Simulation in Abstracted Environments ... 51
4.2.3 Network Analysis and Algorithmic Inference: Structural Diagnostics on Static Landscapes ... 52
4.3 Translational Barriers: CAS Dynamics in Practice ... 53
4.3.1 Path Dependence in Data Infrastructure ... 53
4.3.2 Structural Inertia in Evaluation Metrics ... 54
4.3.3 Co-Evolutionary Mismatches between Computational and Institutional Timelines ... 55
4.4 Synthesis: Translational Barriers as Stable System Features ... 56
4.5 Implications for Architectural Specifications ... 57
References ... 58

**5 Empirical Manifestations of Translational Barriers in Developing Contexts** . . . . . 61
5.1 Introduction: Empirical Evidence of Theoretical Dynamics . . . . . . 61
5.2 Arthur's Path Dependence in Data Infrastructure . . . . . . . . . . . . 61
5.3 Hannan and Freeman's Structural Inertia in Institutional Capacity Development . . . . . . . . . . 63
5.4 Convergent Manifestation: Algorithmic Bias as Systematic Ethical Failure . . . . . . . . . . 66
5.5 Synthesis: The Interdependent System . . . . . . . . . . 68
5.6 Conclusion: Requirements for Complexity-Aware, Ethically Grounded Architecture . . . . . . . . . . 70
References . . . . . . . . . . 71

**6 Empirical Validation: Testing the Complexity-Ethics Framework Through Institutional Practice** . . . . . . . . . . 73
6.1 From Theory to Empirical Testing . . . . . . . . . . 73
6.2 Case Studies: Four Institutional Contexts . . . . . . . . . . 74
6.2.1 UNDP Accelerator Labs and AI Hub: Multilateral Development Architecture . . . . . . . . . . 74
6.2.2 World Bank ESMAP and DE4A: International Financial Institution Architecture . . . . . . . . . . 75
6.2.3 National Energy Efficiency Programs: Domestic Implementation Architecture . . . . . . . . . . 77
6.2.4 ASEAN Centre for Energy: A Theoretical Probe into Regional Coordination . . . . . . . . . . 79
6.3 Architectural Requirements Derived from Empirical Evidence . . . 80
6.4 Empirically Validated Specifications . . . . . . . . . . 83
References . . . . . . . . . . 84

**Part III The BRIDGE-E Framework**

**7 BRIDGE-E: A Complexity-Aware Framework for Policy Coherence** . . . . . . . . . . 89
7.1 From Empirical Requirements to Computational Architecture . . . . 89
7.1.1 Recalling the Empirically Validated Specifications . . . . . . . 89
7.1.2 BRIDGE-E as Operationalization of Established Theory . . . . . . . . . . 90
7.2 BRIDGE-E Architectural Components . . . . . . . . . . 91
7.2.1 Data Harmonization Module: Addressing Path Dependence . . . . . . . . . . 91
7.2.2 Coherence Assessment Engine: Operationalizing Complexity Dynamics and Justice Constraints . . . . . . . . . . 93
7.2.3 Context-Sensitive Recommendation System: Addressing Structural Inertia Through Deliberative Co-Validation . . . . . . . . . . 94

7.3 Integrated Architecture: Cross-Component Synergies . . . 96
7.4 Technical Implementation Specifications . . . 97
7.4.1 Computational Architecture . . . 97
7.4.2 Infrastructure Prerequisites in Resource-Constrained Environments . . . 97
7.4.3 Sustained Viability: Model Maintenance and Adaptation . . . 98
7.5 A Complexity-Aware Approach to Validation and Research . . . 98
7.5.1 From Metrics to Processes: Rethinking Validation . . . 98
7.5.2 Theoretical Boundaries as Generative Research Questions . . . 99
7.5.3 A Transdisciplinary Research Agenda . . . 100
7.6 A Protocol for Implementation: Institutional Primacy in Socio-Technical Design . . . 101
7.6.1 Phase 1: Institutional Diagnostics and Rule Co-Design . . . 101
7.6.2 Phase 2: Collaborative Technical Adaptation and Stewardship Transfer . . . 102
7.7 Potential Impact and Cross-Sectoral Transferability . . . 103
7.8 Conclusion: From Theory to Testable Proposition . . . 103
References . . . 104

**8 Governing Algorithmic Complexity** . . . 109
8.1 The Foundational Challenge: From Implementation Failure to Conceptual Mismatch . . . 109
8.2 The Necessary Evolution: Governing Complexity with Complexity . . . 110
8.3 BRIDGE-E as Theoretical Exemplar: Instantiating Complexity-Aware Governance . . . 112
8.4 The Structural Imperative and the Governance Challenge . . . 113
8.5 Conclusion: From Exemplar to Praxis . . . 114
References . . . 115

**Index** . . . 117

# Part I
# Foundations

In Part I, we establish the analytical foundation. We identify the problem requiring investigation, the theoretical frameworks for understanding it, and the methodology for gathering evidence.

Chapter 1 identifies the governance-data-inequality nexus. We define this nexus as a self-reinforcing pattern where institutional fragmentation produces data infrastructure constraints, which enable algorithmic bias, which legitimizes continued fragmentation. Ultimately, this obstructs holistic policy coherence. Throughout our empirical analysis, institutional capacity deficits, data quality challenges, and algorithmic bias risks emerge as the core components of this governance-data-inequality nexus. We then introduce BRIDGE-E as a novel computational framework addressing these documented barriers in low-data, high-complexity environments.

Chapter 2 develops the complexity-ethics analytical lens. From complexity science, we understand how governance systems resist transformation through technical sophistication alone. We demonstrate how current approaches exhibit complex adaptive systems (CAS) dynamics: Arthur's path dependence, Holland's adaptive systems, Hannan and Freeman's structural inertia, and Kauffman's co-evolutionary dynamics. Fricker's epistemic injustice, Floridi's information ethics, Dignum's responsible AI, and Jenkins et al.'s energy justice help us understand how computational applications perpetuate systematic exclusions. The chapter also examines current applications (e.g., natural language processing, agent-based modeling, network analysis), revealing their characteristic limitations. They all function as diagnostic instruments rather than transformational frameworks.

Chapter 3 details our methodology. We explain why we adopted and adapted the scoping review methodology developed by Arksey and O'Malley. We employed thematic saturation as endpoint, which appeared to be particularly suited to emerging fields with fragmented literatures like computational policy analysis. The chapter addresses epistemic constraints transparently. We acknowledge structural exclusions in academic knowledge production and establish protocols for integrating gray literature.

These three chapters provide the complete conceptual and methodological apparatus for Part II's empirical analysis.

# Part I
# Foundation

# Chapter 1
# Bridging Computational Techniques and Energy Policy Coherence

The Organisation for Economic Co-operation and Development (OECD)'s 2025 analysis *Governing with Artificial Intelligence* (OECD, 2025) provides a nuanced assessment of AI in policy evaluation. The report catalogs concrete AI capabilities with direct relevance for evaluative practice, highlighting two critical dimensions, which are an acceleration of analytical tasks and the enablement of large-scale analysis. OECD documents how AI radically accelerates core tasks, citing evidence that large language models (LLMs) can complete bias assessments of randomized controlled trials in 53 s versus an estimated 28 min per study for human analysts. Furthermore, the report details pioneering implementations at scale, including Norway's performance audit of cybercrime policing using text mining and machine learning (ML), and the OECD's own AI-assisted evaluation of over 1500 environmental policy interventions across 41 countries (OECD, 2025).

The report also notes that "its use within government for policy evaluation has been limited and has progressed slower than in other functions" (OECD, 2025). OECD attributes this constrained adoption to persistent implementation challenges, specifically "lack of high-quality data," "skills gaps," and risks of "automation bias" (OECD, 2025) where human operators may over-rely on algorithmic outputs without sufficient critical scrutiny.

This authoritative diagnosis directly validates the core components of the governance-data-inequality nexus: institutional capacity deficits (skills gaps), informational pathologies (data scarcity and quality), and systemic risks that algorithmic systems may perpetuate rather than mitigate bias. Consequently, the OECD's analysis provides empirical grounding for the BRIDGE-E (Bias-Resistant Integration of Data, Governance, and Equity for Energy) framework's focus on overcoming these documented barriers through explicitly bias-resistant, capacity-sensitive, and equity-centered computational design.

OECD acknowledges that AI capabilities (OECD/UNESCO, 2024) show considerable promise in the public sector although their application has progressed unevenly. AI-driven methods have indeed gained substantial traction in specific



R. Bouderhem, *A Computational Framework for Energy Policy Coherence*, SpringerBriefs in Energy, https://doi.org/10.1007/978-3-032-22978-6_1

technical domains, such as load forecasting, where machine learning (ML) models achieve notable predictive accuracy (Wang et al., 2023). However, methodological frameworks for systematically applying AI to assess policy coherence in developing countries, characterized by data scarcity, institutional fragmentation, and capacity constraints, remain underdeveloped. This implementation gap is particularly consequential given that the very barriers documented by the OECD (data quality, skills, automation risks) are defining features of the governance-data-inequality nexus in developing economies.

The BRIDGE-E framework addresses this gap by proposing a theoretical approach that operationalizes data science capabilities evidenced in the OECD report *Governing with Artificial Intelligence* (OECD, 2025). It uses natural language processing (NLP) for policy document analysis and synthetic data generation for augmenting evidence bases. BRIDGE-E also incorporates specific design safeguards against the documented implementation challenges. These include human-in-the-loop (HITL) validation protocols to mitigate automation bias; data harmonization to address data quality deficits aligning with Findable, Accessible, Interoperable, and Reusable (FAIR) Principles; and capacity-building mechanisms integrated within the framework's deployment model. We position this conceptual framework as a testable design proposition. BRIDGE-E aims to advance the field of computational policy analysis by providing a complexity-aware, ethically grounded approach for policy coherence in precisely those low-data, high-complexity environments where current adoption has lagged.

## 1.1 Background and Problem Statement

The challenge of achieving coherent energy policy is particularly sharp in developing countries, which commonly face fragmented government structures that generate conflicting priorities. This policy incoherence manifests in what Xu has termed an "energy revolution paradox" (Xu, 2021) through their analysis of China's energy transition. Xu analyzes this paradox as a systemic contradiction where support for renewable energy coexists with sustained fossil fuel incentives, leading to outcomes like renewable curtailment alongside fossil expansion. Their research demonstrates that this contradiction arises from socio-institutional factors such as interest divergence and weak policy coordination, rather than from technical limitations (Xu, 2021). This empirical evidence supports our argumentation that the core barrier to effective energy transitions is governance complexity. This fundamental challenge applies to all energy transitions, especially in the Global South.

Governance complexity represents a core characteristic of major energy transitions. A prime example is Germany's *Energiewende*, which is a frequently studied case of deliberate, large-scale transition. Radtke (2025) demonstrate that even in a context of relatively high institutional capacity, sustained progress depends on continuous participatory negotiation, reflexive policy adaptation, and the strategic alignment of initiatives across multiple governance levels. This proves that effective energy transitions are inherently complex socio-political processes requiring adaptive governance mechanisms.

This universal governance complexity resonates forcefully in developing countries. Widespread data limitations—including gaps in coverage, inconsistent formats, and unreliable quality—severely limit the evidence-based coordination needed among different stakeholders. National policies may indeed present unified goals, contrasting with local realities and implementation challenges. This façade reveals a fundamental flaw in relying solely on top-down approaches. This situation challenges the common belief that centralized policy alignment alone can drive sustainability transitions, especially where local institutions vary widely in their capacity and authority.

As established by OECD in its 2018 report on policy coherence for sustainable development (PCSD), the "transformation towards sustainable and resilient societies cannot be achieved without addressing the systemic causes of vulnerability, including inequalities; injustice and discrimination; weak governance and institutions; and the depletion of natural resources" (OECD, 2018). As a response, we design BRIDGE-E to address these interconnected gaps by proposing computational methods tailored to low-data, high-complexity policy environments, with explicit safeguards against algorithmic biases that could marginalize rural or low-income communities.

The BRIDGE-E framework is a computational policy coherence tool leveraging AI and data science methodologies for equitable energy transitions. On the other hand, the European Union (EU) BRIDGE initiative (European Commission, 2023) is a technical blueprint for interoperable energy data exchange in Europe. Their differences lie in scope (policy vs. infrastructure), methodology (AI vs. architecture), and target regions (Global South/rest of the world vs. EU). Together, they represent complementary advances in energy governance and data ecosystems.

Our conceptual framework uses data science methodologies including:

- NLP, knowledge graphs, and synthetic data generation to consolidate disparate data sources for data harmonization.
- Transformer models (BERT), GNNs, and XAI to quantify and interpret policy interactions for coherence assessment.
- HITL systems to ensure that policy suggestions acknowledge local realities and ethical considerations for context-sensitive recommendations.

SDG 7 establishes comprehensive targets to "ensure access to affordable, reliable, sustainable and modern energy for all" by 2030. According to the latest *Tracking SDG 7: The Energy Progress Report* 2025 (IEA et al., 2025), the global energy transition reveals a starkly bifurcated reality. On one hand, installed renewable energy generating capacity reached a record 478 watts per capita globally in 2023 (up from 248 watts in 2015), and international public finance for clean energy in developing countries rose to USD 21.6 billion in 2023, a substantial increase from USD 12.1 in 2015. On the other hand, we have to nuance these promising trends as we observe a profound and concerning persistence in access deficits. Despite the share of the global population with access to electricity reaching almost

92 per cent in 2023, 666 million people still lacked access. Furthermore, the share of renewable energy in total final energy consumption stood at only 17.9% in 2022, while primary energy intensity was 3.87 MJ/USD in 2022 (improved from 4.26 MJ/USD in 2015, but progress too slow).

This alarming stagnation—driven by population growth outpacing connections, the global energy crisis, inflation, and debt distress—is concentrated in Sub-Saharan Africa, where electrification rates remain critically low. Concurrently, 2.1 billion people lack access to clean cooking fuels and technologies in 2023—with major implications for health, gender equality, and the environment. This divergence between technological progress and equitable distribution crystallizes the core challenge. Systemic governance failures are actively undermining the achievement of SDG 7, making the governance-data-inequality nexus not merely an analytical lens but an urgent empirical reality (IEA et al., 2025).

Developing countries encounter multidimensional challenges in implementing effective energy efficiency measures. The industrial sector continues to dominate energy consumption patterns, constituting nearly 40% of total final energy utilization globally (IEA, 2025). At the same time, the buildings and construction sector remains a key driver of the climate crisis, consuming 32% of global final energy and contributing to 34% of global energy-related $CO_2$ emissions (UNEP, 2025). Addressing such inefficiency in developing contexts is uniquely difficult. The implementation of energy policy in developing countries faces profound systemic challenges. These include not only technical and financial constraints such as restricted access to advanced technologies, but also prohibitive capital costs, and insufficient technical capacity. The Global South also faces fundamental failures in governance, where common policy tools like energy subsidies reduce energy access. As shown by Shittu et al. (2024), "fuel subsidies are detrimental to expanding electricity access in developing countries"; strong institutional quality only partially moderates this adverse effect.

Governance frameworks present additional obstacles through inadequate regulatory structures, limited institutional capacity for enforcement, and underdeveloped monitoring and verification systems. Such challenges align with measured deficiencies in institutional transparency and accountability in energy governance worldwide (Drago & Gatto, 2022). Financial market failures compound these difficulties, characterized by insufficient public and private capital allocation. However, research in developing economies suggests well-structured public-private partnerships can help overcome investment barriers, with studies finding positive associations between public-private investment and renewable energy adoption (Yin & Qamruzzaman, 2024).

The COVID-19 pandemic has also disrupted the pathway to energy sustainability, forcing out conventional approaches and necessitating the prioritization of resilient operations aligned with both societal needs and ecosystem health (Elavarasan et al., 2021). This external shock has intensified the cumulative effect of long-standing, interconnected systemic barriers. Research demonstrates that such barriers are interconnected and must be addressed holistically (Chai & Yeo, 2012). Our study identifies these barriers as constituting a governance-data-inequality nexus.

Navigating this reality necessitates integrated analytical approaches and a multi-sectoral coordination. Researchers suggest building sector decarbonization frameworks (Wang & Pan, 2024), designed for developing economies that require new institutional and technological frameworks for energy access (Kaygusuz, 2012). These challenges represent the precise context that the BRIDGE-E framework addresses.

## 1.2 Theoretical Framework and Research Gap

Policy coherence for sustainable development represents a critical theoretical construct for analyzing energy efficiency governance. OECD defines policy coherence as "the systematic promotion of mutually reinforcing policy actions across government departments and agencies creating synergies toward achieving the agreed objectives" (OECD, 2024). Our analytical approach synthesizes three foundational theoretical perspectives that jointly establish the conceptual architecture for computational policy analysis and coherence.

First, we draw upon Holland's emergence (1995, 1998), Arthur's complexity economics (1999), which builds on his seminal work on path dependence and lock-in (Arthur, 1989), and Kauffman's (1993) co-evolutionary dynamics to reconceptualize governance as a complex adaptive system (CAS) rather than mechanistic policy structures amenable to linear optimization. Energy policy environments reveal characteristic features of CAS. This includes emergent behaviors arising from decentralized interactions among heterogeneous actors, positive feedback mechanisms generating path dependencies and institutional lock-in effects, and nonlinear dynamics where small governance interventions can produce disproportionate systemic effects. This complexity perspective challenges conventional assumptions underlying traditional policy analysis methodologies. Complexity science reveals why fragmented governance structures and data scarcity are two sides of the same coin. We cannot fix them through purely technical solutions dissociated from institutional dynamics.

Second, we integrate Fricker's (2007) epistemic injustice theory to analyze how algorithmic systems can systematically marginalize certain forms of knowledge and expertise in policy processes. Fricker identifies two forms of epistemic injustice particularly relevant to computational governance:

1. Testimonial injustice, where credibility deficits prevent certain stakeholders' policy knowledge from receiving appropriate consideration,
2. Hermeneutical injustice, where gaps in collective interpretive resources systematically disadvantage marginalized communities in articulating their energy access experiences.

These concepts provide analytical frameworks for understanding how algorithmic bias in computational policy systems can perpetuate existing inequalities. When we—consciously or not—privilege certain data sources, policy framings, and stakeholder perspectives, we systematically exclude others.

Third, we employ Floridi's (1999, 2013) information ethics and Dignum's (2019) responsible AI paradigm to establish the normative foundations for ethical algorithm design in policy design. Floridi's macroethics frames the entire informational environment (the "infosphere") as a realm of moral concern. Its core injunction is to reduce informational entropy (the degradation, corruption, or obstruction of information) and promote flourishing. We apply this lens to governance as an informational ecosystem, establishing criteria to assess whether algorithmic systems degrade or enrich the epistemic resources available to stakeholder communities. We operationalize then these commitments through Dignum's (2019) accountability-by-design principles (algorithmic transparency, explainability, and human oversight). We argue these principles are nonnegotiable in policy domains affecting vulnerable populations' fundamental rights, such as energy access.

Complexity science reveals systemic interdependencies, epistemic injustice theory exposes knowledge production inequalities, and information ethics establishes normative design criteria. The convergence of these three theoretical perspectives generates the governance-data-inequality nexus as the integrative analytical framework. We demonstrate how institutional fragmentation (governance dimension), data scarcity (information dimension), and algorithmic bias (inequality dimension) operate as mutually constitutive constraints. Fragmented governance structures impede systematic data collection and policy documentation. Data scarcity limits algorithmic model training and validation, increasing susceptibility to bias. Algorithmic systems trained on such contaminated data perpetuate bias, which in turn, leads to governance fragmentation by generating policy recommendations that systematically disadvantage populations already marginalized by existing institutional arrangements.

Previous research has investigated isolated policy instruments for energy efficiency. However, a significant analytical gap remains regarding the systematic application of computational methods to assess policy coherence across multiple sectors and governance levels in developing country contexts. This is precisely where the governance-data-inequality nexus is most acute. The literature predominantly addresses technical applications of AI in energy systems operations such as the potential of AI to reduce energy and carbon emissions of commercial buildings (Ding et al., 2024). Research also focuses on stability analysis and control in smart grids (Shi et al., 2020), sustainable energy (Saheb et al., 2022), demand forecasting, and grid optimization (Xie et al., 2023). However, there is insufficient attention to methodological frameworks for AI deployment in comprehensive policy analysis. Contextual adaptability, equity considerations, and implementation constraints in resource-limited environments are precisely the dimensions that complexity-aware, ethically grounded approaches must address.

Important challenges documented in the literature include the implementation of AI impact assessment (AI-IAs) protocols (Stahl et al., 2023), data scarcity (Amalou et al., 2024), limited technical capacity, and ethical concerns in AI deployment for a sustainable energy industry (Ahmad et al., 2021). These are not isolated technical issues but the constitutive components of the governance-data-inequality nexus we identify. Their combined effect manifests at a global scale, threatening to lock in profound structural disparities. UNDP released an important report in December 2025, which frames the consequence of such barriers as a "Next Great Divergence" (UNDP Asia and the Pacific, 2025). UNDP rightly warns that unequal readiness could set in motion rising inequality between countries, a risk evidenced by stark disparities. AI has reached 1.2 billion users in only 3 years, with nearly 70% in developing countries, yet in some high-income economies 2 in 3 people use AI tools compared to just 5% in many low-income countries.

Confronted with this risk of a "Next Great Divergence," the theoretical contribution of this research lies in its systematic integration of different yet complementary theories. Complexity science, epistemic injustice theory, information ethics, computational social science, and OECD's policy coherence frameworks are the prism under which we frame our framework. We establish a conceptual bridge between theoretical sophistication regarding governance complexity and technological innovation in data science. This approach enables evidence-based policy formulation that addresses both technical capabilities and ethical imperatives for accelerating sustainable energy transitions.

Having established this theoretical landscape, the analysis now requires methodological frameworks capable of addressing these systemic challenges within energy governance contexts. The research objectives articulated below respond directly to the identified gaps, providing operational pathways for investigating computational techniques' capacities to enhance coordination across energy efficiency policies, renewable energy integration strategies, and broader sustainable development objectives.

## 1.3 Research Objectives and Methodological Approach

We examine the role of data science methodologies for policy analysis and coherence in energy efficiency, with a particular attention to progress toward UN SDGs 7 (affordable and clean energy) (UN DESA, n.d.-a), 9 (industry, innovation and infrastructure) (UN DESA, n.d.-b), and 13 (climate action) (UN DESA, n.d.-c). Following a rigorous analysis of existing applications and emergent capabilities, we propose a methodological framework for enhancing policy coherence through the BRIDGE-E framework.

Our research objectives are to:

1. Analyze how machine learning (ML) algorithms (Zekić-Sušac et al., 2021) and natural language processing (NLP) techniques (Smith et al., 2021) can improve the quantitative and qualitative assessment of complex policy landscapes affecting energy efficiency outcomes. We also explore empirical evidence in critical domains such as green buildings (Mahmood et al., 2024).

2. Identify key methodological challenges in computational techniques for policy analysis, including data availability limitations, technical capacity constraints, and ethical considerations related to algorithmic transparency and accountability.
3. Evaluate recent institutional initiatives such as the UNDP AI Hub (UNDP, 2024) as implementation models for context-specific AI deployment in developing countries, with emphasis on marginalized populations and traditional knowledge.
4. Develop a comprehensive research agenda addressing current methodological limitations in AI-enhanced policy analysis for sustainable energy development, our focus being replicability and scalability.

## 1.4 Methodological Framework and Significance

BRIDGE-E represents a conceptual framework, a provocation awaiting empirical validation. However, its design draws on documented patterns of energy policy implementation failures in developing countries. Institutional fragmentation, inadequate financing mechanisms, and governance coordination deficits systematically impede equitable energy transitions. Kenya's renewable energy strategy exemplifies these challenges (Rotich et al., 2024). Despite achieving approximately 90% renewable electricity generation capacity and establishing ambitious targets for 100% renewable energy transition by 2020, the country failed to meet this goal, with only 76.89% of the population accessing electricity by 2022 (Rotich et al., 2024). Documented barriers include poor policy implementation, inadequate financing, infrastructure deficits, outdated policies, overambitious project design, and high operational costs. These implementation failures disproportionately affected rural populations already experiencing severe energy access inequalities, with 86% biomass dependence in rural areas compared to 21% in urban areas, and approximately 75% of rural populations lacking electricity access compared to 23% in urban areas (Rotich et al., 2024). This urban-rural disparity exemplifies how governance failures led to differential implementation outcomes.

This empirical evidence reveals systemic barriers that constitute the governance-data-inequality nexus. The lack of robust data collection systems, inadequate mechanisms for policy coherence, and weak institutional coordination perpetuate existing inequalities. Bias-aware synthetic data generation would address data scarcity constraints. Decentralized governance principles would overcome institutional fragmentation, and human-in-the-loop validation protocols would ensure context-sensitive implementations. This is the objective of the BRIDGE-E framework.

Our complexity-ethics analytical lens helps us identify these governance failures. The persistent urban-rural disparity and institutional stagnation exemplify Arthur's (1989, 1999) path dependence and lock-in, where early infrastructure investments create self-reinforcing trajectories that exclude rural areas. This also manifests Fricker's (2007) dual epistemic injustice. Rural populations suffer testimonial injustice as the computational systems discount their energy needs in policy prioritization. Additionally, rural populations suffer hermeneutical injustice as policy frameworks lack the conceptual resources to articulate rural electrification challenges distinct from urban grid-extension models. From Floridi's (1999, 2013) perspective, this represents informational entropy. Precisely, we observe a degradation of the governance "infosphere" caused by data scarcity, weak coordination, and incoherent policy, which actively perpetuates inequality.

The BRIDGE-E framework counteracts these specific theoretical pathologies through three integrated design principles, each operationalizing our foundational frameworks:

1. Countering path dependence and informational entropy via epistemic inclusion: to break Arthur's (1989) lock-in and reduce Floridi's (2013) informational entropy, BRIDGE-E employs bias-aware synthetic data generation. This addresses data scarcity not by reinforcing existing (urban-biased) data pathways but by creating community-validated data to represent excluded realities, expanding hermeneutical resources.
2. Overcoming fragmentation through complexity-aware coordination: to navigate the complex adaptive system of governance, BRIDGE-E incorporates decentralized coordination principles derived from Holland's (1995, 1998) insights on emergence and adaptation. We combine these principles with Arthur's (1989, 1999) analysis of path dependence and lock-in, and Kauffman's (1993) co-evolutionary dynamics. By doing so, we aim to overcome institutional silos by modeling and facilitating policy coherence across jurisdictions.
3. Ensuring accountability and context-sensitivity via responsible AI: to operationalize Dignum's (2019) accountability-by-design and mitigate Fricker's (2007) testimonial injustice, BRIDGE-E implements human-in-the-loop (HITL) validation protocols. This embeds continuous stakeholder oversight, ensuring local knowledge and ethical considerations are not just technical optimization, but inherent to algorithmic recommendations.

We also use the OECD's policy coherence framework to define the practical targets for this approach, which combines computation with solid theory.

Employing a systematic scoping review methodology, we analyzed publicly available data sources from peer-reviewed publications, institutional reports, and policy documents. Our methodological approach draws upon Arksey and O'Malley's (2005) foundational five-stage framework, encompassing research question identification, relevant study identification, study selection, data charting, and results synthesis. This foundational work established scoping studies as systematic instruments for mapping research activity and clarifying key concepts in fields lacking synthesized evidence bases. This foundation is complemented by contemporary guidance from the Joanna Briggs Institute (Peters et al., 2020), which provides updated protocols addressing iterative search refinement, quality appraisal considerations, and evidence synthesis approaches appropriate to diverse methodological contexts in nascent interdisciplinary domains.

The methodological framework enabled the systematic synthesis and mapping of technical capabilities, implementation barriers, and institutional success factors across diverse geographical and governance environments. This structured synthesis of existing evidence guided the development of our theoretical framework. We maintained a transparent distinction between documented evidence patterns and conceptual propositions requiring future empirical validation.

Our analytical framework incorporates both technological dimensions (i.e., algorithm selection, data requirements, system architecture) and implementation factors (i.e., institutional capacity, ethical considerations, contextual adaptation). Such an approach provides a comprehensive assessment of computational applications in this domain. This comprehensive method enables the identification of success factors and potential barriers to effective implementation.

We believe this study will make a significant contribution to the emerging field of data science methodologies for advancing sustainable development goals by establishing methodological foundations for quantitative assessment of policy coherence. It highlights both the analytical potential of AI models in predicting multidimensional policy impacts and the necessity of context-specific, ethically grounded approaches to implementation. By integrating AI-driven data harmonization, policy coherence assessment, and context-sensitive recommendations, BRIDGE-E tackles fragmented governance, data scarcity, and algorithmic bias, all documented barriers to equitable energy transitions. The findings also outline a structured research agenda that addresses current analytical limitations while supporting evidence-based policymaking for energy efficiency in developing countries, thus advancing progress toward SDG 7 targets through enhanced policy.

We conceptualize this relationship not as a linear process but as a reinforcing loop. The articulation of all components illustrates how computational capabilities directly tackle the fundamental barriers of fragmentation, data scarcity, and bias. However, the feedback loop is equally critical. The implementation of AI-generated, context-sensitive policy interventions alters the real-world governance and energy landscape. This conceptual framework generates new, and ideally more coherent, streams of policy documentation, implementation data, and energy consumption metrics. As a result, these novel data serve as a refined training corpus for the AI models, enabling continuous learning, model refinement, and adaptation to evolving conditions. This closed-loop system positions BRIDGE-E not as a static assessment tool but as an adaptive learning instrument, capable of improving its own performance and relevance over time through interaction with the policy environment it seeks to improve.

As we identified the empirical landscape of energy policy fragmentation, it is now essential to discuss policy coherence frameworks. Contradictory policy instruments, implementation gaps between governance levels, and insufficient cross-sectoral alignment demonstrate the need to develop new analytical approaches to policy coherence. These approaches must be capable of managing complex interdependencies characteristic of contemporary energy systems. Chapter 2 evaluates the traditional concept of policy coherence and identifies its limitations for computational implementation. In response, we develop the complexity-ethics analytical lens necessary for navigating these challenges.

## References


Ahmad, T., Zhang, D., Huang, C., Zhang, H., Dai, N., Song, Y., & Chen, H. (2021). Artificial intelligence in sustainable energy industry: Status quo, challenges and opportunities. *Journal of Cleaner Production, 289*, 125834. https://doi.org/10.1016/j.jclepro.2021.125834

Amalou, I., Mouhni, N., Abdali, A., Jakimi, A., & Tourad, M. C. (2024). Deep learning for smart grid application: Addressing data scarcity challenges and enhancing load forecasting efficiency. In Y. M. Elhadj, M. F. Nanne, A. Koubaa, F. Meziane, & M. Deriche (Eds.), *Artificial intelligence and its practical applications in the digital economy* (Lecture notes in networks and systems) (Vol. 861). Springer. https://doi.org/10.1007/978-3-031-71426-9_2

Arksey, H., & O'Malley, L. (2005). Scoping studies: Toward a methodological framework. *International Journal of Social Research Methodology, 8*(1), 19–32. https://doi.org/10.1080/1364557032000119616

Arthur, W. B. (1989). Competing technologies, increasing returns, and lock-in by historical events. *The Economic Journal, 99*(394), 116–131. https://doi.org/10.2307/2234208

Arthur, W. B. (1999). Complexity and the economy. *Science, 284*, 107–109. https://doi.org/10.1126/science.284.5411.107

Chai, K.-H., & Yeo, C. (2012). Overcoming energy efficiency barriers through systems approach—A conceptual framework. *Energy Policy, 46*, 460–472. https://doi.org/10.1016/j.enpol.2012.04.012

Dignum, V. (2019). *Responsible artificial intelligence: How to develop and use AI in a responsible way*. Springer. https://doi.org/10.1007/978-3-030-30371-6

Ding, C., Ke, J., Levine, M., et al. (2024). Potential of artificial intelligence in reducing energy and carbon emissions of commercial buildings at scale. *Nature Communications, 15*, 5916. https://doi.org/10.1038/s41467-024-50088-4

Drago, C., & Gatto, A. (2022). Policy, regulation effectiveness, and sustainability in the energy sector: A worldwide interval-based composite indicator. *Energy Policy, 167*, 112889. https://doi.org/10.1016/j.enpol.2022.112889

Elavarasan, R. M., Pugazhendhi, R., Jamal, T., Dyduch, J., Arif, M. T., Kumar, N. M., Shafiullah, G. M., Chopra, S. S., & Nadarajah, M. (2021). Envisioning the UN sustainable development goals (SDGs) through the lens of energy sustainability (SDG 7) in the post-COVID-19 world. *Applied Energy, 292*, 116665. https://doi.org/10.1016/j.apenergy.2021.116665

European Commission: Directorate-General for Energy, Couto, M. J., Kotsalos, K., & Kukk, K. (2023). *European (energy) data exchange reference architecture 3.0*. Publications Office of the European Union. https://data.europa.eu/doi/10.2833/815043

Floridi, L. (1999). Information ethics: On the philosophical foundation of computer ethics. *Ethics and Information Technology, 1*, 33–52. https://doi.org/10.1023/A:1010018611096

Floridi, L. (2013). *The ethics of information*. Oxford University Press. https://doi.org/10.1093/acprof:oso/9780199641321.001.0001

Fricker, M. (2007). *Epistemic injustice: Power and the ethics of knowing*. Oxford University Press. https://doi.org/10.1093/acprof:oso/9780198237907.001.0001

Holland, J. H. (1995). *Hidden order: How adaptation builds complexity*. Addison-Wesley.

Holland, J. H. (1998). *Emergence: From chaos to order*. Oxford University Press. https://doi.org/10.1093/oso/9780198504092.001.0001

IEA. (2025). *Energy efficiency 2025*. IEA. https://www.iea.org/reports/energy-efficiency-2025 (Licence: CC BY 4.0)

IEA, IRENA, UNSD, WHO, & World Bank. (2025). *Tracking SDG 7: The energy Progress report 2025*. World Bank Group. https://trackingsdg7.esmap.org/downloads

Kauffman, S. A. (1993). *The origins of order: Self-organization and selection in evolution*. Oxford Academic. https://doi.org/10.1093/oso/9780195079517.001.0001

Kaygusuz, K. (2012). Energy for sustainable development: A case of developing countries. *Renewable and Sustainable Energy Reviews, 16*(2), 1116–1126. https://doi.org/10.1016/j.rser.2011.11.013

Mahmood, S., Sun, H., El-kenawy, E. S. M., et al. (2024). Integrating machine and deep learning technologies in green buildings for enhanced energy efficiency and environmental sustainability. *Scientific Reports, 14*, 20331. https://doi.org/10.1038/s41598-024-70519-y

OECD. (2018). *Policy coherence for sustainable development 2018: Toward sustainable and resilient societies*. OECD Publishing. https://doi.org/10.1787/9789264301061-en

OECD. (2024). *Ensuring policy coherence through better policy development, co-ordination and stakeholder participation in Romania* (OECD public governance policy papers, no. 46). OECD Publishing. https://doi.org/10.1787/a9d50cc9-en

OECD. (2025). *Governing with artificial intelligence: The state of play and way forward in Core government functions*. OECD Publishing. https://doi.org/10.1787/795de142-en

OECD/UNESCO. (2024). *G7 toolkit for artificial intelligence in the public sector*. OECD Publishing. https://doi.org/10.1787/421c1244-en

Peters, M. D. J., Marnie, C., Tricco, A. C., Pollock, D., Munn, Z., Alexander, L., McInerney, P., Godfrey, C. M., & Khalil, H. (2020). Updated methodological guidance for the conduct of scoping reviews. *JBI Evidence Synthesis, 18*(10), 2119–2126. https://doi.org/10.11124/JBIES-20-00167

Radtke, J. (2025). Understanding the complexity of governing energy transitions: Introducing an integrated approach of policy and transition perspectives. *Environmental Policy and Governance, 35*(4), 595–614. https://doi.org/10.1002/eet.2158

Rotich, I. K., Chepkirui, H., & Musyimi, P. K. (2024). Renewable energy status and uptake in Kenya. *Energy Strategy Reviews, 54*, 101453. https://doi.org/10.1016/j.esr.2024.101453

Saheb, T., Dehghani, M., & Saheb, T. (2022). Artificial intelligence for sustainable energy: A contextual topic modeling and content analysis. *Sustainable Computing: Informatics and Systems, 35*, 100699. https://doi.org/10.1016/j.suscom.2022.100699

Shi, Z., Yao, W., Li, Z., Zeng, L., Zhao, Y., Zhang, R., Tang, Y., & Wen, J. (2020). Artificial intelligence techniques for stability analysis and control in smart grids: Methodologies, applications, challenges and future directions. *Applied Energy, 278*, 115733. https://doi.org/10.1016/j.apenergy.2020.115733

Shittu, I., Saqib, A., Abdul Latiff, A. R., & Baharudin, S. A. (2024). Energy subsidies and energy access in developing countries: Does institutional quality matter? *SAGE Open, 14*(3). https://doi.org/10.1177/21582440241271118

Smith, T. B., Vacca, R., Mantegazza, L., et al. (2021). Natural language processing and network analysis provide novel insights on policy and scientific discourse around Sustainable Development Goals. *Scientific Reports, 11*, 22427. https://doi.org/10.1038/s41598-021-01801-6

Stahl, B. C., Antoniou, J., Bhalla, N., et al. (2023). A systematic review of artificial intelligence impact assessments. *Artificial Intelligence Review, 56*, 12799–12831. https://doi.org/10.1007/s10462-023-10420-8

UN DESA. (n.d.-a). *Sustainable Development. Goal 7: Ensure access to affordable, reliable, sustainable and modern energy for all*. https://sdgs.un.org/goals/goal7

UN DESA. (n.d.-b). *Sustainable Development. Goal 9: Build resilient infrastructure, promote inclusive and sustainable industrialization and foster innovation*. https://sdgs.un.org/goals/goal9

UN DESA. (n.d.-c). *Sustainable Development. Goal 13: Take urgent action to combat climate change and its impacts*. https://sdgs.un.org/goals/goal13

UNDP. (2024, July 17). *G7 Italia, AI hub for sustainable development: Strengthening local AI ecosystems through collective action*. https://www.undp.org/sites/g/files/zskgke326/files/2024-07/ai_hub_report_digital.pdf

UNDP Asia and the Pacific. (2025, December 2). *The next great divergence: Why AI may widen inequality between countries*. https://www.undp.org/asia-pacific/next-great-divergence

UNEP. (2025, March 17). *Global status report for buildings and construction 2024/2025*. UN Environment Programme. https://wedocs.unep.org/handle/20.500.11822/47214

Wang, Y., & Pan, W. (2024). Multi-sectoral efforts are required for decarbonising the building sector: A case in Hong Kong. *NPJ Urban Sustainability, 4*, 43. https://doi.org/10.1038/s42949-024-00180-x

Wang, X., Wang, H., Bhandari, B., et al. (2023). AI-empowered methods for smart energy consumption: A review of load forecasting, anomaly detection and demand response. *International Journal of Precision Engineering and Manufacturing-Green Technology, 11*, 963–993. https://doi.org/10.1007/s40684-023-00537-0

Xie, L., Zheng, X., Sun, Y., Huang, T., & Bruton, T. (2023). Massively digitized power grid: Opportunities and challenges of use-inspired AI. *Proceedings of the IEEE, 111*(7), 762–787. https://doi.org/10.1109/JPROC.2022.3175070

Xu, S. (2021). The paradox of the energy revolution in China: A socio-technical transition perspective. *Renewable and Sustainable Energy Reviews, 137*, 110469. https://doi.org/10.1016/j.rser.2020.110469

Yin, C., & Qamruzzaman, M. (2024). Empowering renewable energy consumption through public-private investment, urbanization, and globalization: Evidence from CS-ARDL and NARDL. *Heliyon, 10*(4), e26455. https://doi.org/10.1016/j.heliyon.2024.e26455

Zekić-Sušac, M., Mitrović, S., & Has, A. (2021). Machine learning based system for managing energy efficiency of public sector as an approach towards smart cities. *International Journal of Information Management, 58*, 102074. https://doi.org/10.1016/j.ijinfomgt.2020.102074

# Chapter 2
# Complexity Science and Ethical Frameworks for Computational Governance

Energy governance operates within a landscape of profound theoretical complexity. Institutional structures, normative frameworks, and computational methodologies intersect in ways that demand rigorous conceptual grounding. Here, we establish the foundations necessary for understanding how computational methods can meaningfully contribute to policy analysis and coherence. We also acknowledge the epistemological and ethical constraints that govern such applications.

The conceptualization of policy coherence has evolved considerably over the last two decades. The doctrine moved beyond simplistic notions of alignment toward sophisticated understandings of systemic interdependence across different fields. For developing countries pursuing SDG 7, the challenge extends beyond mere technical coordination. It encompasses fundamental questions about knowledge production, institutional capacity, and equitable governance that require multidimensional analytical frameworks and significant institutional and technical resources.

This chapter proceeds through five interconnected analytical dimensions. Section 2.1 examines the conceptual ambiguity and political reality of policy coherence. Section 2.2 develops an integrated analytical lens combining complexity science and ethical foundations. Section 2.3 diagnoses the epistemological challenges and translational gap that hinder computational implementation. Section 2.4 analyzes the Organisation for Economic Co-operation and Development (OECD)'s operational policy coherence framework as an implementation benchmark. Finally, Sect. 2.5 explores the current state of computational applications in policy analysis.

Together, these theoretical perspectives provide the conceptual architecture for the empirical investigation that follows. They reveal that effective computational approaches to policy coherence must integrate technical sophistication with institutional sensitivity and ethical accountability. This represents a synthesis that conventional governance frameworks have yet to achieve. The BRIDGE-E framework is our proposition.



R. Bouderhem, *A Computational Framework for Energy Policy Coherence*, SpringerBriefs in Energy, https://doi.org/10.1007/978-3-032-22978-6_2

## 2.1 Policy Coherence: Conceptual Ambiguity and Political Reality

In 2006, May et al. argued that "the concept [of policy coherence] is under-theorized and has received little empirical examination" (May et al., 2006). Nothing has changed since then, or very little. Recent research suggests that policy coherence "remains an elusive concept that refers to and includes multiple analytical dimensions… Its meaning has been variously defined and unsystematically explored" (Righettini & Lizzi, 2022).

Different definitions emerged for this concept. Nilsson et al. (2012) define policy coherence as "an attribute of policy that systematically reduces conflicts and promotes synergies between and within different policy areas to achieve the outcomes associated with jointly agreed policy objectives." Policy coherence is indeed a concept that we cannot easily grasp. Some researchers even consider that "perfect policy coherence is unattainable and may be ill-advised" (Cairney, 2025).

Righettini and Lizzi (2022) demonstrate that researchers commonly address policy coherence under the prism of sustainable development. In addition, they note that policy coherence has become an important point of debate since 2015 within international organizations such as the United Nations, OECD, WTO, or WHO (Righettini & Lizzi, 2022). This also coincides with the launch of the 2030 Sustainable Development Agenda in 2015 (Righettini & Lizzi, 2022).

The literature recognizes the complex interplay between various policy domains in sustainable development (Shawoo et al., 2023). In this diverse field, researchers apprehend this complexity through different analytical lenses such as the institutional approach (Ashoff, 2005), the policy integration approach (Tosun & Lang, 2017), and the systems thinking approach (Castro, 2022).

A recent study, however, reframes the core challenge of policy coherence. Dzebo et al. (2025) note that policy coherence is central to sustainable development, but its "conceptual ambiguity and political complexity limit its effectiveness." They note that "policy coherence tends to be conceptualized as technocratic and depoliticized. Tools such as PCD [policy coherence for development] and PCSD [policy coherence for sustainable development], in particular, have been criticized for creating an illusion that policy trade-offs and conflicts are problems to be solved with more effective management. Such framing, in turn, prevents open and meaningful discussions about deeper political and social conflicts that are inherent in policymaking" (Dzebo et al., 2025).

To overcome this, Dzebo et al. (2025) introduce a necessary political lens and examine coherence through the interplay of "ideas, institutions, and interest." They demonstrate that "power dynamics and competing values fundamentally shape outcomes." They conclude that policy coherence is a normative process. It requires transparent value choices about "coherence against what," and documents that "incoherence produces unintended distributional effects, often burdening vulnerable populations and benefiting elites" (Dzebo et al., 2025).

This synthesis allows us to identify a critical gap for computational policy analysis. Dzebo et al. (2025) show that effective policy coherence necessitates navigating political complexity, normative choices, and equity impacts. We argue that current computational tools deployed are inherently designed for qualitative, human-scale deliberation. They lack the capacity required for governance in complex and data-scarce environments.

Advancing policy coherence for sustainable development and equitable energy transitions requires a new theoretical lens that can bridge this gap. This lens must guide computational methodologies capable of modeling dynamic complexity and account for diverse ethical and political dimensions.

Section 2.2 develops this essential foundation by integrating complexity science with frameworks of epistemic justice, information ethics, responsible AI, and energy justice. This approach establishes the conceptual basis for the BRIDGE-E framework.

## 2.2 An Integrated Analytical Lens: Complexity, Ethics, and the Computational Challenge

### *2.2.1 Complexity Science, Economics, and the Theory of Complex Adaptive Systems (CAS)*

Path dependence (Arthur, 1989), emergence (Holland, 1995, 1998), and co-evolution (Kauffman, 1993) offer the explanatory mechanisms for understanding how current systems generate the specific pathologies of fragmentation and inertia in energy governance.

Where traditional policy models assume equilibrium and linear causality, complexity science reveals governance as a complex adaptive system (CAS). Governance is indeed a dynamic network where institutions, policies, and stakeholders co-evolve in ways that generate emergent, often unpredictable outcomes (Vivo et al., 2024). This perspective, grounded in seminal work from the Santa Fe Institute by Arthur (1989), Kauffman (1993) and Holland (1995, 1998), helps explain why our governance-data-inequality nexus persists as a self-reinforcing pathology. The fragmentation we observe emerges from three fundamental complexity dynamics.

First, institutional fragmentation exhibits structural inertia and resistance to change. This resistance is explained by path dependence, an economic concept formalized by Arthur (1989). It shows how an early choice in a system becomes locked-in because shifting away from it later becomes too complex and costly. This could be the creation of separate energy and transport departments for example. The classic historical case is David's analysis of the QWERTY keyboard, which became standard due to early adoption, not superior design (David, 1985). When this pattern of lock-in occurs in public organizations, it creates what Hannan and Freeman (1984) called structural inertia. This describes how an organization's core structure resists change because maintaining reliable operations and clear accountability often conflicts with adapting to new circumstances. Consequently, siloed energy governance is not a simple mistake. It is a stable, inefficient socio-technical lock-in.

Such a persistent state is the direct consequence of the combination of economic path dependence and organizational structural inertia.

Second, policy interventions fail due to co-evolution within dynamic fitness landscapes. A policy's efficacy is not intrinsic but determined by its adaptive fit within a co-evolving system of regulations, markets, and social norms. This environment constitutes a dynamic fitness landscape (Wright, 1932). Kauffman (1993) later formalized this concept for complex adaptive systems, while Levinthal (1996) dealt with dynamic fitness landscapes for organizational contexts in his analysis of learning and evolutionary competitive dynamics.

Their work demonstrates how interdependencies create rugged landscapes characterized by a dynamic of adaptation to changing conditions. Consequently, successful policy transfer fails because each governance context presents a unique, evolving topography. This explains why we need to evaluate interventions within the governance-data-inequality nexus as context-specific adaptations, not standardized solutions (e.g., urban-rural divide, participation of all stakeholders in the decision-making process especially marginalized and Indigenous populations, local specificities such as data scarcity, implementation challenges).

Third, system-level pathologies like fragmentation are emergent properties. These macroscopic patterns arise from bottom-up interactions of heterogeneous agents (e.g., ministries, firms, citizens) following local rules. This represents a defining feature of complex adaptive systems (CAS). As Holland (1995, 1998) established, we cannot reduce such emergent patterns to component analysis. This creates an epistemological mandate for generative computational science (Epstein, 2007), employing methods like agent-based modeling to simulate the interaction dynamics that produce systemic pathologies. The BRIDGE-E framework's architecture responds directly to this methodological imperative.

Understanding energy governance through this tripartite complexity lens—path-dependent lock-in, co-evolution on dynamic landscapes, and non-deductible emergence—reveals why traditional, reductionist policy approaches consistently underperform. They attempt to impose linear order on nonlinear complexity. The BRIDGE-E framework emerges from this theoretical foundation as a computational approach designed to navigate these complexity challenges. We draw insights from both foundational complexity science (Mitchell, 2009) and contemporary applications to governance (Kolt et al., 2025).

### 2.2.2 Ethical Foundations: From Epistemic Justice to Responsible AI

The development of computational policy frameworks requires grounding in complementary ethical traditions that address both social and technical dimensions of governance. Fricker's (2007) theory of epistemic injustice provides tools for understanding how knowledge systems can systematically exclude certain perspectives.

Testimonial injustice "occurs when prejudice causes a hearer to give a deflated level of credibility to a speaker's word," while hermeneutical injustice arises "when a gap in collective interpretive resources puts someone at an unfair disadvantage when it comes to making sense of their social experiences" (Fricker, 2007).

Floridi's information ethics (2013) treats the entire information environment ("infosphere") as the focus of moral concern. This framework emphasizes that technology should promote human flourishing by reducing entropy (disorder, destruction, and corruption) in the information environment. For policy systems, this means designing computational tools that enhance coherence, transparency, and accessibility of governance information.

Dignum's (2019) model of responsible AI emphasizes responsibility-by-design. As such, it requires embedding moral reasoning capabilities directly into AI systems rather than treating ethics as an external constraint. This approach advocates for AI systems to be transparent, accountable, and capable of reasoning about norms, values, and potential consequences of their actions.

These combined frameworks provide a comprehensive ethical foundation. To ensure a comprehensive application to the energy sector, we anchor these ethical principles in the established framework of energy justice developed by Jenkins et al. (2016). Its three principles directly guide our commitment to equity within energy governance. Distributional justice ensures an equitable allocation of benefits and burdens. Recognition justice demands respect for diverse communities and perspectives. Procedural justice focuses on inclusive and fair decision-making.

Here, Fricker's (2007) epistemic injustice helps us diagnose the failures in recognition and procedure, Floridi's (2013) information ethics provides the teleological direction for a just "infosphere," and Dignum (2019) offers the methodological approach for implementation. Path dependence, emergence, and co-evolution explain the mechanisms through which governance systems generate the structural biases, hermeneutical gaps, and informational entropy that these ethical frameworks seek to redress.

This integrated lens—combining complexity science's explanatory power with ethics' normative guidance—provides the analytical tools to assess traditional policy coherence theory through a computational implementation perspective. Section 2.3 diagnoses the epistemological challenges and translational barriers that hinder computational implementation.

## 2.3 Epistemological Challenge and Translational Barriers

Contemporary approaches to policy coherence emphasize the need to grapple with complex policy environments, often using integrative modeling (Collste et al., 2017). This marks a shift from earlier, linear models that assumed straightforward cause-and-effect relationships (Flanagan et al., 2011). Recent frameworks explicitly adopt a complexity perspective (Serban, 2024). This tripartite evolution of policy coherence theories—from organizational structures to integrated objectives and finally to systemic interactions—reveals a fundamental epistemological challenge. If coherence emerges from systemic interactions and diverse perspectives, it cannot be reduced to a technical checklist. Policy coherence is not an objective per se, or a measurable state through calculated metrics. It is an interpretive construct, intrinsic to stakeholder perspectives, political priorities, and the specific socio-technical system in question. According to Dzebo et al. (2025), policy coherence is an inherently political process. It is shaped by power and competing values, and its common technocratic framing obscures this reality.

This interpretive nature of policy coherence exemplifies what measurement theory describes as an "unobservable theoretical construct" (Jacobs & Wallach, 2021). It cannot be measured directly but must be operationalized through necessarily imperfect proxies. Such a process introduces the potential for significant mismatches between the theoretical understanding of the construct and its measurement (Jacobs & Wallach, 2021). These theoretical approaches provide conceptual lenses for understanding policy coherence. However, they demonstrate significant resistance to computational implementation. We identify such challenges as translational barriers that themselves demonstrate the CAS dynamics these theories describe.

These translational challenges arise from the complexity dynamics that policy coherence theories identify. Arthur's path dependence (1989) explains how early technological choices become locked-in due to switching costs. Here, early theoretical design choices in policy coherence frameworks create barriers to subsequent computational translation. Similarly, just as Hannan and Freeman's structural inertia (1984) explains organizational resistance to change, these interpretive frameworks resist reduction to the deterministic, formal logic required for algorithms. The reliance on qualitative interpretation and context-dependence necessarily amplifies these obstacles in resource-constrained environments.

These translational barriers are not merely technical shortcomings but symptoms of deeper, pre-existing systemic pathologies in policy systems. These are captured in Weber and Rohracher's "failures framework" (2012), which distinguishes between structural and, more critically, transformational system failures. These include directionality failure (absence of strategic vision), policy coordination failure (institutional fragmentation), and reflexivity failure (inadequate capacity for learning). A computational system, such as an AI model, trained on data and documents from a governance environment suffering from these failures does not create incoherence anew; it inherits, scales, and algorithmically codifies it. The translational impedance occurs when formal logic encounters a system already structured to resist coherence.

The profound risk inherent in this dynamic is what Fricker (2007) considers a systematic "wrong done to someone specifically in their capacity as a knower." In policy systems, this manifests in two key forms:

1. Testimonial injustice: the credibility deficit given to the testimony of marginalized stakeholders whose lived experiences are discounted in policy analysis.

2. Hermeneutical injustice: the structural absence of conceptual resources to adequately understand and communicate the social experiences of disadvantaged groups.

When AI systems are trained on data that reflects these existing power imbalances, they do not correct for epistemic injustice; they automate and scale it. The algorithmic bias observed in urban-centric energy models is thus not a technical flaw but a digital manifestation of this deeper epistemic failure.

This hermeneutical gap becomes computationally consequential when policy data schemas and AI training corpora lack the ontological richness to represent marginalized realities. For instance, energy policy databases may categorize energy access through binary grid-connection metrics. This makes invisible the complex energy landscapes of informal settlements that rely on layered, precarious energy sources. The resulting AI models cannot see what their conceptual framework cannot name. This explains why merely adding more data to biased systems often fails. What is required is expanding the hermeneutical capacity of the computational system itself. This need aligns with emerging work in digital hermeneutics, which examines how digital systems interpret and are interpreted (Romele et al., 2020).

These translational challenges highlight a core problem within computational social science (Lazer et al., 2009). Arthur's (1989) lock-in and DiMaggio and Powell's (1983) institutional isomorphism elucidate the persistence of existing governance systems. However, the application of these same CAS dynamics to the translation process itself reveals why new computational methodologies struggle to model the complexity of those systems. Translating a multi-dimensional construct such as policy coherence into computable inputs requires a framework that does not seek a single "ground truth" (Bowker & Star, 2000). Instead, such a framework should model and reconcile multiple, often competing perspectives.

This epistemological reality is a first-order principle for the BRIDGE-E framework. We aim to make the conflicting dimensions of coherence explicit and negotiable, rather than imposing a false, automated consensus. This requires architectural innovations such as the reinforcement learning and adaptive governance components integrated into the BRIDGE-E framework to address the gap between theoretical comprehension and computational application. The BRIDGE-E framework, therefore, is not merely an application of AI to policy, but a conceptual bridge designed to make these nuanced social and political realities computationally tractable.

These CAS-driven translational barriers crystallize the core implementation challenge. There is a significant gap between theoretical comprehension and algorithmic application. Now that we have identified this challenge, we need concrete specifications for what constitutes a successful implementation. Section 2.4 examines the OECD's Policy Coherence for Sustainable Development (PCSD) framework as an institutional benchmark for our framework. OECD PCSD appears to be the essential performance criteria for the deployment of any computational tool for coherence.

## 2.4 OECD Approach: Policy Coherence for Sustainable Development (PCSD)

Policy coherence is a foundational principle for addressing complex, interconnected challenges such as achieving energy efficiency within the broader framework of the Sustainable Development Goals (SDGs). OECD has developed a comprehensive and authoritative framework for Policy Coherence for Sustainable Development (PCSD).

### *2.4.1 Core Definition and Purpose*

The OECD's formal legal definition of Policy Coherence for Sustainable Development (PCSD), as stated in the Recommendation of the Council on Policy Coherence for Sustainable Development, Section I (OECD, 2019a), is:

> [A]n approach to integrate the dimensions of sustainable development throughout domestic and international policy-making. Its objectives in the context of the 2030 Agenda are to advance the integrated implementation of the 2030 Agenda by: (i) Fostering synergies and maximising benefits across economic, social and environmental policy areas; (ii) Balancing domestic policy objectives with internationally recognised sustainable development goals; and (iii) Addressing the transboundary and long-term impacts of policies, including those likely to affect developing countries.

According to another seminal report released by OECD in 2018, this framework is a direct operational response to SDG Target 17.14. OECD calls on all countries to "enhance policy coherence for sustainable development" as a key means of implementation (OECD, 2018).

### *2.4.2 The Eight Institutional Building Blocks*

The OECD 2019 report, *Policy Coherence for Sustainable Development 2019: Empowering People and Ensuring Inclusiveness and Equality*, explains that the ability to implement coherent policies depends on "institutional mechanisms, i.e. processes, structures and working methods, used by governments to manage and co-ordinate policy" (OECD, 2019b). It identifies eight key building blocks:

1. Political commitment and leadership
2. Long-term vision and planning horizons
3. Policy integration
4. Policy and institutional coordination

5. Regional and local involvement
6. Stakeholder engagement
7. Analysis and assessments of policy impacts
8. Monitoring, reporting and evaluation.

In this 2019 report, OECD emphasizes that "there is no one-size-fits-all approach to improve PCSD" (OECD, 2019b). Each country must tailor these mechanisms to its context.

### 2.4.3 *The 2019 Council Recommendation: A Three-Pillar Structure for Action*

The OECD Council Recommendation on PCSD (2019a) translates these analytical building blocks into formal, actionable guidance for governments. It structures its recommendations under three pillars as shown in Table 2.1.

### 2.4.4 *Practical Application: Evidence from Romania*

The OECD's 2024 policy paper on Romania applies the PCSD lens, finding that "the strategic planning apparatus would benefit from providing a more cohesive narrative around strategic priorities due to a fragmented strategic framework" (OECD, 2024). The assessment identifies coherence challenges and recommends tools—such as standardizing interministerial committees and strengthening e-consultation—that align directly with the OECD's principles for improving policy development, coordination, and stakeholder engagement.

**Table 2.1** The three-pillar framework for Policy Coherence for Sustainable Development (PCSD), as established by the OECD Council Recommendation (2019a)

| Pillar | Objective | Key guidance for adherents |
|---|---|---|
| I. Strategic vision | To achieve the 2030 agenda in an integrated manner | Build political commitment, define a long-term vision, and improve policy integration (e.g., using the budget process) |
| II. Institutional mechanisms | To address policy interactions across sectors and government levels | Ensure whole-of-government coordination, engage sub-national levels, and involve stakeholders proactively |
| III. Responsive tools | To anticipate, assess, and address policy impacts | Implement policy impact analyses and strengthen monitoring, reporting, and evaluation systems |

### 2.4.5 *Implications for Energy Efficiency (SDG 7)*

Applying the OECD PCSD framework is critical for energy efficiency, a cross-cutting goal. The OECD's 2018 report states a core principle: "the SDGs cannot be achieved through single-sector or silo approaches" (OECD, 2018). It examines SDG 7 (Affordable and Clean Energy) alongside other goals, highlighting that policy integration is essential to manage synergies and trade-offs, for example, between energy, water, and land use. This aligns with recommendations made by UNDP for facilitating integration and coherence for SDG implementation (UNDP, 2017).

For developing countries, coherent energy efficiency policies must therefore:

- Establish a strategic vision: integrate efficiency targets into national development plans with stable, long-term policy signals.
- Strengthen institutional mechanisms: coordinate actions across relevant ministries (e.g., energy, finance, industry, transport) and engage local authorities in implementation.
- Utilize responsive tools: conduct ex-ante impact assessments to ensure policies do not adversely affect vulnerable groups and establish robust monitoring systems to track progress.

The OECD PCSD framework specifies the institutional requirements ("*what*") of coherent governance. This specification raises a pivotal technological question for the BRIDGE-E framework: do existing computational tools possess the capability to meet these requirements? Do they augment or transform the "*how*" of implementation?

The OECD PCSD framework establishes a test for computational policy analysis. Section 2.5 interrogates the state of AI applications against this test. We demonstrate that current computational tools remain primarily diagnostic instruments that cannot autonomously enable the integrated vision, coordination, and adaptation that coherent governance requires. This crystallizes the imperative for a new approach.

## 2.5 Computational Applications in Policy Analysis: Capabilities and Limitations

Computational methodologies across the policy cycle are relatively recent. A decade ago, researchers already praised their promise to enhance analytical capacity (Höchtl et al., 2016). This section evaluates contemporary applications through the complexity-ethics lens developed in Sect. 2.2. Our analysis reveals a consistent pattern. Current computational tools function primarily as sophisticated diagnostic instruments that operate on governance systems but remain external to their core adaptive dynamics. This pattern emerged from a foundational tension recognized early in the field: a "tremendous potential" alongside significant risks to "statistical bias, privacy, equity, and governance" (Schintler & Kulkarni, 2014). These tools provide critical insights yet collectively demonstrate the translational barriers diagnosed in Sect. 2.3. This reveals a fundamental gap between formal logic and the co-evolutionary reality (Kauffman, 1993) of policy systems.

### *2.5.1 Concrete Illustrations for Energy Efficiency (SDG 7)*

Current applications reveal a pattern of specialized diagnostic tools, each offering powerful capabilities while exhibiting constitutive limitations when viewed through the prism of complex adaptive systems (CAS).

**Natural Language Processing (NLP): Pattern Recognition Without Co-Evolution**
NLP enables large-scale analysis of policy texts. This data science methodology quantifies discursive patterns such as gendered language in energy plans (Carroll et al., 2024) or thematic clusters in AI strategies (Saheb et al., 2022). However, NLP treats language as a static dataset. It cannot model policy discourse as a strategic, co-evolving instrument that shapes and is shaped by institutional interests within dynamic fitness landscapes (Kauffman, 1993; Levinthal, 1996). Recent research confirmed this limitation. Even advanced large language models (LLMs) cannot independently generate policy briefings judged "useful" by experts, failing at the interpretive synthesis central to policy work (Safaei & Longo, 2024).

**Agent-Based Modeling (ABM): Simulating Emergence, Abstracting Structure**
ABM excels at showing how decentralized interactions (e.g., household energy choices) generate emergent macro-level outcomes (Rai & Henry, 2016). In doing so, it exemplifies Holland's (1995, 1998) principle of emergence. ABM disregards the path-dependent institutional structures (legal frameworks, bureaucratic routines, political economies) that form the rigid topology of real-world implementation. ABM simulates possible behaviors but cannot adjudicate what historically constituted systems, characterized by regulatory lock-in (Arthur, 1989) and organizational inertia (Hannan & Freeman, 1984), will permit.

**Network Analysis: Structural Diagnostics on Static Landscapes**
Network analysis maps interdependencies and trade-offs, such as between Sustainable Development Goals (Swain & Ranganathan, 2021) and water-energy-food policies (Suda et al., 2024). It provides unparalleled structural diagnostics. However, it produces a static snapshot of a dynamic fitness landscape (Kauffman, 1993). As Suda et al.'s (2024) findings demonstrate, identifying incoherence is analytically and politically distinct from resolving it; resolution requires navigating the co-evolutionary pressures and power asymmetries that a static diagram cannot simulate.

### 2.5.2 Synthesis: Translational Barriers in Current Applications

These computational applications reveal a consistent limitation. They recognize patterns, simulate abstracted dynamics, and map static structures, but all these functions observe the system from the outside.

When evaluated through the complexity lens, they systematically address certain system dimensions such as emergence and structure. However, they do not account for emergence (Holland, 1995, 1998), path dependence (Arthur, 1989), structural inertia (Hannan & Freeman, 1984), and co-evolutionary dynamics (Kauffman, 1993; Levinthal, 1996); these patterns explain policy failures. These systems can only diagnose the pathologies of fragmentation from a position that cannot alter the system's deeper logic. This represents the empirical manifestation of the translational barriers identified in Sect. 2.3.

Emerging hybrid approaches combining AI analysis with expert validation acknowledge this gap but seek to address it through task combination. For example, Carić et al.'s (2026) six-phase framework for policy coherence achieves strong metrics (Fleiss' $\kappa = 0.717$) but confirms the limitation. They acknowledge that "expert oversight remains indispensable for contextual interpretation, ethical validation, and policy relevance" (Carić et al., 2026).

Empirical research validates this limitation. A recent study found that data science and machine learning (ML) are functionally limited to data-driven governance where decision-making responsibilities are low, such as strategic assessments or resource allocation (Kim, 2024). Their application to core evidence-based regulation is constrained by a lack of local specificity, susceptibility to environmental uncertainty, and an inability to provide causal explanations. This empirical finding validates the analysis presented in Sect. 2.3. Even the most advanced tools cannot make the kind of contextual, causally complex, and high-stakes decisions that are essential for policy-making.

### 2.5.3 The Architectural Imperative

The challenge shifts from refining diagnostic tools to meeting an architectural imperative. The unmet need is for governed meta-frameworks that plan human-computational interaction within the complex adaptive system of governance. The OECD PCSD framework (Sect. 2.4) specifies the institutional requirements (vision, coordination, engagement, adaptation) that current applications support but cannot fulfill autonomously. This defines the critical gap. There is an urgent need to move from tools that operate within path-dependent systems to frameworks that can help transform their co-evolutionary dynamics. This imperative directly motivates the BRIDGE-E framework, developed in Chap. 7, which proposes to build a complexity-aware and ethically grounded adaptive logic into the core of computational policy analysis.

Chapter 3 establishes the methodological protocol for the comprehensive scoping review that systematically maps this empirical landscape, identifying the precise contours of the translational challenges this chapter has diagnosed.

## References

Arthur, W. B. (1989). Competing technologies, increasing returns, and lock-in by historical events. *The Economic Journal, 99*(394), 116–131. https://doi.org/10.2307/2234208

Ashoff, G. (2005). *Enhancing policy coherence for development: Justification, recognition and approaches to achievement (No. 11)*. Studies. https://www.econstor.eu/bitstream/10419/199136/1/die-study-11.pdf

Bowker, G. C., & Star, S. L. (2000). *Sorting things out: Classification and its consequences*. MIT Press.

Cairney, P. (2025). Why perfect policy coherence is unattainable (and may be ill-advised). *Policy Sciences, 58*(3), 619–642. https://doi.org/10.1007/s11077-025-09582-9

Carić, H., Mandić, A., & Sever, I. (2026). A six-phase AI-expert framework for evaluating policy coherence in sustainable tourism. *Information Technology & Tourism, 28*, 6. https://doi.org/10.1007/s40558-025-00352-0

Carroll, P., Singh, B., & Mangina, E. (2024). Uncovering gender dimensions in energy policy using natural language processing. *Renewable and Sustainable Energy Reviews, 193*, 114281. https://doi.org/10.1016/j.rser.2024.114281

Castro, C. V. (2022). Systems-thinking for environmental policy coherence: Stakeholder knowledge, fuzzy logic, and causal reasoning. *Environmental Science & Policy, 136*, 413–427. https://doi.org/10.1016/j.envsci.2022.07.001

Collste, D., Pedercini, M., & Cornell, S. E. (2017). Policy coherence to achieve the SDGs: Using integrated simulation models to assess effective policies. *Sustainability Science, 12*, 921–931. https://doi.org/10.1007/s11625-017-0457-x

David, P. A. (1985). Clio and the economics of QWERTY. *The American Economic Review, 75*(2), 332–337.

Dignum, V. (2019). *Responsible artificial intelligence: How to develop and use AI in a responsible way*. Springer. https://doi.org/10.1007/978-3-030-30371-6

DiMaggio, P. J., & Powell, W. W. (1983). The iron cage revisited: Institutional isomorphism and collective rationality in organizational fields. *American Sociological Review, 48*(2), 147–160. https://doi.org/10.2307/2095101

Dzebo, A., Shawoo, Z., & Browne, K. (2025). Does policy coherence make National implementation of global sustainability agendas more successful? *Annual Review of Environment and Resources, 50*. https://doi.org/10.1146/annurev-environ-111523-102337

Epstein, J. M. (2007). *Generative Social Science*. Princeton University Press. https://doi.org/10.1515/9781400842872

Flanagan, K., Uyarra, E., & Laranja, M. (2011). Reconceptualising the 'policy mix' for innovation. *Research Policy, 40*(5), 702–713. https://doi.org/10.1016/j.respol.2011.02.005

Floridi, L. (2013). *The ethics of information*. Oxford University Press. https://doi.org/10.1093/acprof:oso/9780199641321.001.0001

Fricker, M. (2007). *Epistemic injustice: Power and the ethics of knowing*. Oxford University Press. https://doi.org/10.1093/acprof:oso/9780198237907.001.0001

Hannan, M. T., & Freeman, J. (1984). Structural inertia and organizational change. *American Sociological Review, 49*(2), 149–164. https://doi.org/10.2307/2095567

Höchtl, J., Parycek, P., & Schöllhammer, R. (2016). Big data in the policy cycle: Policy decision making in the digital era. *Journal of Organizational Computing and Electronic Commerce, 26*(1–2), 147–169. https://doi.org/10.1080/10919392.2015.1125187
Holland, J. H. (1995). *Hidden order: How adaptation builds complexity*. Addison-Wesley
Holland, J. H. (1998). *Emergence: From chaos to order*. Oxford Academic. https://doi.org/10.1093/oso/9780198504092.001.0001
Jacobs, A. Z., & Wallach, H. (2021, March). Measurement and fairness. In *Proceedings of the 2021 ACM conference on fairness, accountability, and transparency* (pp. 375–385). https://doi.org/10.1145/3442188.3445901.
Jenkins, K., McCauley, D., Heffron, R., Stephan, H., & Rehner, R. (2016). Energy justice: A conceptual review. *Energy Research & Social Science, 11*, 174–182. https://doi.org/10.1016/j.erss.2015.10.004
Kauffman, S. A. (1993). *The origins of order: Self-organization and selection in evolution*. Oxford Academic. https://doi.org/10.1093/oso/9780195079517.001.0001
Kim, E. S. (2024). Can data science achieve the ideal of evidence-based decision-making in environmental regulation? *Technology in Society, 78*, 102615. https://doi.org/10.1016/j.techsoc.2024.102615
Kolt, N., Shur-Ofry, M., & Cohen, R. (2025). Lessons from complex systems science for AI governance. *Patterns, 6*(8), 101341. https://doi.org/10.1016/j.patter.2025.101341
Lazer, D., Pentland, A., Adamic, L., Aral, S., Barabási, A. L., Brewer, D., et al. (2009). Computational social science. *Science, 323*(5915), 721–723. https://doi.org/10.1126/science.1167742
Levinthal, D. (1996). Learning and Schumpeterian dynamics. In G. Dosi & F. Malerba (Eds.), *Organization and strategy in the evolution of the Enterprise*. Palgrave Macmillan. https://doi.org/10.1007/978-1-349-13389-5_2
May, P. J., Sapotichne, J., & Workman, S. (2006). Policy coherence and policy domains. *Policy Studies Journal, 34*, 381–403. https://doi.org/10.1111/j.1541-0072.2006.00178.x
Mitchell, M. (2009). *Complexity: A guided tour*. Oxford Academic. https://doi.org/10.1093/oso/9780195124415.001.0001
Nilsson, M., Zamparutti, T., Petersen, J. E., Nykvist, B., Rudberg, P., & McGuinn, J. (2012). Understanding policy coherence: Analytical framework and examples of sector–environment policy interactions in the EU. *Environmental Policy and Governance, 22*, 395–423. https://doi.org/10.1002/eet.1589
OECD. (2018). *Policy coherence for sustainable development 2018: Towards sustainable and resilient societies*. OECD Publishing. https://doi.org/10.1787/9789264301061-en
OECD. (2019a). *Recommendation of the council on policy coherence for sustainable development*. OECD/LEGAL/0381, adopted on: 29/04/2010, amended on: 11/12/2019. https://legalinstruments.oecd.org/en/instruments/oecd-legal-0381
OECD. (2019b). *Policy coherence for sustainable development 2019: Empowering people and ensuring inclusiveness and equality*. OECD Publishing. https://doi.org/10.1787/a90f851f-en
OECD. (2024). *Ensuring policy coherence through better policy development, co-ordination and stakeholder participation in Romania* (OECD public governance policy papers, no. 46). OECD Publishing. https://doi.org/10.1787/a9d50cc9-en
Rai, V., & Henry, A. (2016). Agent-based modelling of consumer energy choices. *Nature Climate Change, 6*, 556–562. https://doi.org/10.1038/nclimate2967
Righettini, M. S., & Lizzi, R. (2022). How scholars break down "policy coherence": The impact of sustainable development global agendas on academic literature. *Environmental Policy and Governance, 32*(2), 98–109. https://doi.org/10.1002/eet.1966
Romele, A., Severo, M., & Furia, P. (2020). Digital hermeneutics: From interpreting with machines to interpretational machines. *AI & Society, 35*, 73–86. https://doi.org/10.1007/s00146-018-0856-2
Safaei, M., & Longo, J. (2024). The end of the policy analyst? Testing the capability of artificial intelligence to generate plausible, persuasive, and useful policy analysis. *Digital Government: Research and Practice, 5*(1), 1–35. https://doi.org/10.1145/3604570

Saheb, T., Dehghani, M., & Saheb, T. (2022). Artificial intelligence for sustainable energy: A contextual topic modeling and content analysis. *Sustainable Computing: Informatics and Systems, 35*, 100699. https://doi.org/10.1016/j.suscom.2022.100699

Schintler, L. A., & Kulkarni, R. (2014). Big data for policy. *Review of Policy Research, 31*, 343–348. https://doi.org/10.1111/ropr.12079

Serban, I. D. (2024). Wicked problems and policy coherence in global Britain: Lessons learned from the UK context on achieving the SDGs. In *Policy capacity, design and the sustainable development goals: Wicked problems in uncertain environments* (pp. 241–257). Emerald Publishing Limited. https://doi.org/10.1108/978-1-80455-686-320241021

Shawoo, Z., Maltais, A., Dzebo, A., & Pickering, J. (2023). Political drivers of policy coherence for sustainable development: An analytical framework. *Environmental Policy and Governance, 33*(4), 339–350. https://doi.org/10.1002/eet.2039

Suda, A. O., Sušnik, J., Masia, S., & Jewitt, G. (2024). Policy coherence assessment of water, energy, and food resources policies in the Tana River Basin, Kenya. *Environmental Science & Policy, 159*, 103816. https://doi.org/10.1016/j.envsci.2024.103816

Swain, R. B., & Ranganathan, S. (2021). Modeling interlinkages between sustainable development goals using network analysis. *World Development, 138*, 105136. https://doi.org/10.1016/j.worlddev.2020.105136

Tosun, J., & Lang, A. (2017). Policy integration: Mapping the different concepts. *Policy Studies, 38*(6), 553–570. https://doi.org/10.1080/01442872.2017.1339239

UNDP. (2017). *Institutional and coordination mechanisms: Guidance note on facilitating integration and coherence for SDG implementation.* https://sustainabledevelopment.un.org/content/documents/2478Institutional_Coordination_Mechanisms_GuidanceNote.pdf

Vivo, P., Katz, D. M., & Ruhl, J. B. (2024). A complexity science approach to law and governance. *Philosophical Transactions of the Royal Society A, 382*(2270), 20230166. https://doi.org/10.1098/rsta.2023.0166

Weber, K. M., & Rohracher, H. (2012). Legitimizing research, technology and innovation policies for transformative change: Combining insights from innovation systems and multi-level perspective in a comprehensive 'failures' framework. *Research Policy, 41*(6), 1037–1047. https://doi.org/10.1016/j.respol.2011.10.015

Wright, S. (1932). The roles of mutation, inbreeding, crossbreeding and selection in evolution. In *Proceedings of the 6th international congress of genetics* (Vol. 1, pp. 356–366).

# Chapter 3
# An Adapted Scoping Review Protocol for Complex Fields

This chapter details the scoping review methodology (Arksey & O'Malley, 2005) employed to examine AI applications in energy policy, with a particular emphasis on implementation contexts characterized by constrained institutional capacity, deficient data infrastructure, and critical equity imperatives. Additionally, we took into consideration the role played by AI and digitization in achieving SDG 7 targets (Vinuesa et al., 2020) and affordable, clean energy (Awogbemi et al., 2024).

## 3.1 Methodological Rationale

A fundamental epistemological challenge inherent to emerging interdisciplinary fields guided our methodology. How do we assess dispersed and heterogeneous knowledge across fields that have yet to achieve consensus on core constructs, definitions, analytical methodologies, or evaluative criteria? This is the case for policy coherence and computational policy analysis as explained in Chap. 2.

Relevant scholarship is fragmented across legal theory, computer science (focusing on algorithmic optimization), public administration (emphasizing governance mechanisms), development studies (prioritizing equity considerations), and energy systems analysis (addressing technical efficiency). All these fields demonstrate divergent epistemological traditions.

Consequently, we need to prioritize methodologies that maintain not only scientific rigor but also adaptability. This approach ensures we address adequately different fields where knowledge boundaries emerge iteratively rather than through predetermined taxonomic structures.

We employ here a scoping review methodology. Our goal is to survey the breadth of research, map its key concepts, and identify critical knowledge gaps in the emerging field of computational energy policy analysis. We follow the well-established scoping review methodology by Arksey and O'Malley (2005). The protocol



R. Bouderhem, *A Computational Framework for Energy Policy Coherence*, SpringerBriefs in Energy, https://doi.org/10.1007/978-3-032-22978-6_3

encompasses five stages: defining research questions, identifying relevant studies, selecting sources, charting the data, and synthesizing findings. The relevance of this methodology is significant for mapping research where evidence is diverse and not yet consolidated. We have also incorporated refinements from Levac et al. (2010), which emphasize an iterative approach to question development and searching beyond standards databases. Finally, we applied the updated guidance from the Joanna Briggs Institute (Peters et al., 2020). This updated methodology provides protocols for optimizing search strategies for varied types of evidence, and extracting data from the heterogeneous sources typical of nascent, interdisciplinary fields.

Munn et al. (2018) provide a definitive framework for selecting review methods based on research objectives and evidence characteristics. They compare systematic and scoping reviews and demonstrate that a scoping review is the appropriate methodological choice when the research aims for broad coverage across diverse evidence types (Munn et al., 2018). Scoping reviews are also a strategic choice for conceptual mapping, and the identification of research gaps in emerging fields (Munn et al., 2018). These objectives are distinct from those of a traditional systematic review. Contrary to scoping reviews, systematic reviews focus on the depth of evidence, critical appraisal of study quality, and the quantitative aggregation of results to address well-defined policy questions with established outcome measures.

The PRISMA Extension for Scoping Reviews (PRISMA-ScR), developed by Tricco et al. (2018), sets current reporting standards through 20 required and two optional items designed for scoping reviews. A full PRISMA-ScR checklist is beyond the scope of this study but our methodological documentation adheres to its core principles of transparency. This includes identifying evidence sources, stating explicit eligibility criteria, detailing the selection process, following structured data extraction, and fully acknowledging limitations. This structured approach ensures not only the flexibility to investigate rigorously emerging fields defined by conceptual diversity, evolving terminology, and evidence scattered across disconnected yet complementary disciplines. The subsequent Sect. 3.2 details the systematic scoping review protocols.

## 3.2 Systematic Scoping Review: Design and Implementation

We structured our scoping review around five core stages: (1) formulation of iterative research questions, (2) execution of comprehensive multi-source searches, (3) application of explicit inclusion/exclusion criteria, (4) structured data extraction and quality assessment, and (5) synthesis guided by the principle of thematic saturation.

As mentioned, the review followed the enhanced methodological framework for scoping studies (Arksey & O'Malley, 2005; Levac et al., 2010) and aligns with contemporary guidance from the Joanna Briggs Institute (Peters et al., 2020). A defining and adaptive feature of our protocol is the operationalization of thematic saturation as the endpoint for corpus finalization. This approach, adapted from qualitative research methodologies (Saunders et al., 2018; Hennink et al., 2017), is specifically suited to interdisciplinary fields where the research frontier is defined by conceptual emergence rather than by a fixed population of studies.

Our framework for saturation assessment operates across three convergent dimensions:

- Technical saturation: the point at which new sources cease to reveal novel computational methodologies, algorithms, or architectures not already represented in the corpus.
- Contextual saturation: the point at which additional case studies or implementations no longer surface governance barriers, institutional factors, or equity challenges substantively different from established patterns.
- Theoretical saturation: the point at which new literature fails to generate conceptual insights requiring modification of the analytical framework or research questions.

We considered we reached saturation when monitoring across these three dimensions indicated diminishing informational returns. This method prioritizes conceptual adequacy and pattern robustness—essential for mapping a fragmented domain—over an unattainable and mechanically exhaustive enumeration. We detail this process for assessing this convergence in the selection protocol (Sect. 3.2.4).

### 3.2.1 Research Questions

The systematic scoping review addresses one primary research question and three secondary analytical questions. We developed our research questions through an iterative process. Our initial focus was on listing how computational tools are applied. We then shifted to a deeper investigation of the systemic barriers that prevent these tools from effectively addressing the interconnected challenges of governance, data, and inequality. Finally, we focused our inquiry on the solutions proposed to overcome these barriers.

Our primary research focus:

What evidence exists on the use of computational tools for energy policy analysis in developing countries? What does this evidence reveal about the main challenges and potential solutions for making these tools work in practice?

Three secondary research questions complement our primary focus:

RQ1: What are the main technical, institutional, and ethical challenges in using computational methods for energy policy in developing countries?

RQ2: How do governance, data, and equity interact in real-world case studies across different institutional settings?

RQ3: What computational solutions or design principles are proposed to overcome these challenges in resource-constrained environments?

A scoping review is not just suitable but essential for this study. This methodological choice matches the characteristics of our research:

1. Ontological and epistemological heterogeneity: We need to review and examine evidence from fundamentally different types of knowledge and domains (technical algorithm studies, policy evaluation reports, and ethical theory). Scoping reviews can handle this diversity without treating one type of knowledge as more valid than another (Munn et al., 2018).
2. Exploratory synthesis for theory development: Our first step is to survey how ideas connect and identify the depth and breadth of data. This creates a foundational map, which is essential before we can build deeper theoretical arguments in later chapters. A systematic review, which aims to answer a focused "what works" question, would be premature and unsuitable for this exploratory goal (Peters et al., 2020).
3. Alignment with domain complexity: Analyzing how computational tools interact with complex governance systems is by nature unpredictable and adaptive (Arthur, 1989; Kauffman, 1993; Holland, 1995). A rigid, predetermined research method would clash with this reality. The scoping review methodology, which adapts as the research unfolds and stops when no new major themes emerge (thematic saturation), provides the flexibility needed to engage with this kind of complexity (Saunders et al., 2018).

The scoping review methodology is not a default choice. It is a deliberate, epistemologically grounded selection that establishes the evidence base upon which the subsequent theoretical analysis in Chaps. 4 and 5 will be performed. In doing so, we use the complexity-ethics framework from Chap. 2 to analyze the identified patterns as manifestations of path dependence (Arthur, 1989), structural inertia (Hannan and Freeman, 1984), co-evolutionary dynamics (Kauffman, 1993), emergence (Holland, 1995), and epistemic injustice (Fricker, 2007). Chapter 6 will then subject this diagnosis to empirical validation through institutional case studies. This sequenced analysis culminates in the empirically informed, theoretically grounded specifications for the BRIDGE-E framework presented in Chap. 7.

### *3.2.2 Search Strategy and Database Selection*

We developed a comprehensive search strategy using carefully selected keyword combinations designed to capture the intersection of different fields: data science methodologies, policy analysis and coherence, energy efficiency, and sustainable development. Table 3.1 presents the primary keyword combinations employed in this study.

**Table 3.1** Primary keyword combinations for literature search

| Primary term | Secondary term | Tertiary term |
|---|---|---|
| "artificial intelligence" | "policy coherence" | "energy efficiency" |
| "machine learning/ML" | "policy analysis" | "SDG 7" |
| "natural language processing/NLP" | "energy policy" | "developing countries" |
| "AI" | "policy integration" | "clean energy access" |
| "deep learning" | "policy frameworks" | "sustainable energy" |
| "predictive analytics" | "policy evaluation" | "energy transition" |
| "computational methods" | "regulatory impact" | "energy governance" |

**Table 3.2** Information sources for literature review

| Category | Database/Resource | Justification |
|---|---|---|
| Academic databases | Web of science | Comprehensive coverage of peer-reviewed literature across disciplines |
| | Scopus | Broad coverage of technical and policy journals |
| | Google scholar | Access to both academic and gray literature |
| | IEEE Xplore | Specialized coverage of technical AI applications |
| Specialized resources | International energy agency's data and statistics portal | Authoritative source for energy policy and statistics |
| | UN sustainable development goals database | Official repository for SDG-related publications |
| | UNDP AI hub documentation | Specific focus on AI applications for sustainable development |
| | OECD policy Coherence for sustainable development publications | Specialized coverage of policy coherence frameworks |

The selection of keywords was an iterative process designed to capture the multidisciplinary nature of the field. Initial scoping searches revealed that overly broad terms generated unmanageable noise, while overly narrow terms missed critical studies at the intersections of AI, governance, and development. The final combinations in Table 3.1 represent a balance between recall and precision.

Additionally, the temporal boundary of the search to 2015–2025 publications was intentional as we decided to focus on the modern era of computational methods. The current era is indeed characterized by the rise of applied AI, generative AI (GenAI), machine learning (ML), and deep learning, which are fundamentally different from the expert systems that dominated earlier periods. This methodical approach ensures the review's relevance to current technological capabilities and their application to the post-2015 SDG framework.

To ensure comprehensive coverage, we utilized both academic databases and specialized resources, as detailed in Table 3.2.

### 3.2.3 Inclusion and Exclusion Criteria

Clear inclusion and exclusion criteria ensure the relevance and quality of the reviewed studies. Table 3.3 presents the detailed criteria used for literature selection.

For inclusion of gray literature, we defined "high-quality institutions" as those meeting at least three of the following criteria:

- Recognized international organizations with established expertise in energy or sustainability such as OECD, UN agencies, EU, and IEA.
- Government agencies at national or multinational level.
- Research institutions with documented records in computational methods, energy policy, or sustainable development.
- Organizations with transparent methodologies and peer-review processes for publications.
- Entities with established data validation procedures.

### 3.2.4 Selection Protocol and Quality Assessment Procedures

The field is both new and interdisciplinary. Therefore, we used systematic methods to select the strongest studies. Our focus was on identifying and synthesizing the most methodologically robust research on the deployment of computational tools for energy policy analysis and coherence.

1. Systematic Search Strategy

**Table 3.3** Inclusion and exclusion criteria

| Inclusion Criteria | Exclusion Criteria |
|---|---|
| Studies focusing on computational applications in policy analysis related to energy efficiency | Studies not addressing computational methods (e.g., AI, ML, NLP, etc.), policy development, policy analysis and coherence, or energy efficiency; exclusion of the expert systems' era in data science |
| Research addressing policy development, analysis, and coherence in the context of SDG 7 | Publications before 2015 (pre-SDG framework adoption) |
| Peer-reviewed journal articles and conference proceedings | Non-peer-reviewed sources (with exceptions for specified institutional reports) |
| Reports from specified high-quality institutions (UN agencies, OECD, IEA, World Bank, major research institutes) | Opinion pieces, blog posts, news articles |
| Publications in English | Non-English publications without available translations |
| Studies providing explicit methodological details on AI applications | Conceptual papers without practical applications or empirical evidence |

As detailed in Table 3.2, we conducted comprehensive searches across four major academic databases (Web of Science, Scopus, IEEE Xplore, and Google Scholar). We also reviewed specialized institutional repositories (IEA Data and Statistics, UN SDG Global Database, UNDP AI Hub, OECD Policy Coherence for Sustainable Development publications) using the keyword combinations detailed in Table 3.1. As mentioned, the temporal scope (2015–2025), captures the modern era of machine learning (ML) applications and the post-2015 SDG framework.

2. Selection Protocol

We screened literature against the inclusion and exclusion criteria specified in Table 3.3. Studies were required to demonstrate: (1) explicit AI or computational methodology applications, (2) direct relevance to policy coherence or energy efficiency analysis, (3) connection to SDG 7 contexts or developing country implementations, and (4) sufficient methodological transparency to enable critical evaluation.

The selection process incorporated citation network analysis, which consists of a systematic examination of reference lists and forward citation tracking. The goal was to identify seminal contributions that might not surface through keyword searches alone due to terminology evolution across disciplines. This approach aligns with established scoping review protocols that emphasize comprehensiveness through multiple identification pathways as per Levac et al.'s (2010) methodology.

3. Quality Assessment

All retained literature underwent rigorous evaluation using the quality assessment framework detailed in Table 3.4. We included studies achieving a minimum average score of 3.0 across methodological rigor, energy governance relevance, and computational application in the final analytical corpus.

For gray literature from institutional sources (e.g., UN agencies, OECD, World Bank, IEA), supplementary credibility criteria were applied: (1) recognized institutional expertise in energy or development domains, (2) transparent methodological documentation, (3) established data validation procedures, and (4) peer review or external validation processes where applicable.

When selecting studies, we resolved potential discrepancies through reevaluation, with final decisions anchored in the predefined criteria of Table 3.4.

4. Operationalizing Thematic Saturation

Thematic saturation served as the methodological endpoint for corpus finalization. We assessed saturation through monitoring three analytical dimensions:

1. Technical saturation: the point at which successive literature additions ceased revealing novel computational methodologies (e.g., algorithms, architectures, and preprocessing techniques) beyond those already documented in the corpus.
2. Contextual saturation: the point at which additional case studies no longer revealed implementation barriers, institutional factors, or governance challenges substantively different from patterns already identified across geographical contexts and governance levels.

**Table 3.4** Quality assessment criteria

| Dimension | Assessment criteria | Rating scale |
|---|---|---|
| Methodological rigor | Clear research objectives | 1–5 |
| | Appropriate methodology for stated objectives | 1–5 |
| | Transparency in data collection and analysis | 1–5 |
| | Validation procedures | 1–5 |
| | Limitations acknowledged | 1–5 |
| Relevance to energy governance and SDG 7 | Explicit connection to policy coherence frameworks | 1–5 |
| | Clear linkage to SDG 7 targets/indicators | 1–5 |
| | Practical applications for policy development | 1–5 |
| | Consideration of developing country contexts | 1–5 |
| Computational application description | Technical details of AI methods | 1–5 |
| | Data requirements clearly specified | 1–5 |
| | Implementation considerations addressed | 1–5 |
| | Evaluation of AI performance | 1–5 |
| | Ethical considerations discussed | 1–5 |

3. Theoretical saturation: the point at which new literature ceased generating conceptual insights requiring modification of the analytical framework or research question refinement.

We determined saturation when these three dimensions converged. This convergence indicated that the corpus provided sufficient empirical and conceptual coverage to address the research questions articulated in Sect. 3.2.1.

Our method adapts the principle of saturation from qualitative research (Saunders et al., 2018; Hennink et al., 2017) for use in a literature review. In new, interdisciplinary fields, saturation does not mean finding every possible source. Instead, it means we have gathered enough evidence to understand the key concepts and themes.

5. Data Extraction Framework

Our final collection of sources was broad enough to support strong conclusions. It included a wide range of technical tools (e.g., natural language processing, machine learning), covered multiple regions (such as Sub-Saharan Africa and Southeast Asia), and examined different levels of governance (international, regional, national). This breadth ensures reliability and allows us to identify patterns relevant to our research questions developed in Sect. 3.2.1.

We used a structured framework (Table 3.5) to extract data from each study. This ensured we systematically captured key details like publication information, technical methods, policy areas, coherence metrics, relevance to energy goals (SDG 7), implementation contexts, and main findings.

**Table 3.5** Data extraction fields

| Category | Extraction fields |
|---|---|
| Publication information | Authors, year, title, journal/source, publication type, country/region of focus |
| AI techniques | ML algorithms (e.g., random forests, neural networks), NLP methods (e.g., sentiment analysis, topic modeling), other computational approaches |
| Policy domains | Energy efficiency standards, renewable energy integration, building codes, industrial policies, transportation efficiency |
| Coherence assessment | Metrics used to assess policy coherence, analytical frameworks, evaluation methodology |
| SDG 7 relevance | Specific targets addressed, indicators used, alignment with SDG framework |
| Implementation context | Developing/developed country focus, regional considerations, institutional factors |
| Methodological details | Data sources, sample sizes, validation approaches |
| Challenges and limitations | Technical challenges, data issues, ethical considerations, implementation barriers |
| Key findings | Main outcomes related to policy coherence, AI performance, SDG 7 progress |

6. Reflexivity and Selection Bias Mitigation

Our use of thematic saturation to decide when to stop gathering sources introduces an interpretive element. We acknowledge this could reflect our own perspectives. To reduce this risk, we applied four safeguards.

First, we used a standard quality checklist (Table 3.4) to evaluate all literature objectively, minimizing subjective judgment.

Second, we defined explicit, structured criteria for saturation across three areas (technical methods, implementation contexts, and theoretical frameworks) to create clear, consistent stopping points.

Third, we analyzed citation networks to identify influential studies that the research community itself recognizes. This ensures we included foundational work that keyword searches might miss.

Fourth, we added reports from major institutions (e.g., UNDP, IEA, OECD) to capture practical insights often missing from peer-reviewed journals, which tend to focus more on theory.

Together, these steps help ensure our final collection of studies accurately represents the research field, not just our own assumptions.

7. Researcher Perspective

We conducted this review using an interdisciplinary approach. We drew from legal studies, computational governance, energy policy, and AI ethics. This broad perspective helped us connect separate bodies of literature, though it also means our interpretation is influenced by this specific set of disciplines.

We designed our analysis to balance technical computational capabilities, real-world governance challenges, and ethical concerns. This reflects the same integrated approach behind the BRIDGE-E framework. As noted by Tricco et al. (2018), such a positionality statement is important because systematic scoping reviews involve interpretive choices shaped by the researchers' theoretical perspective. This is particularly true in the interdisciplinary fields of computational policy analysis.

### 3.2.5 Synthesis and Thematic Analysis

Our synthesis used a hybrid thematic analysis. We started with a deductive framework based on our research questions developed in Sect. 3.2.1 and the theories from Part I (covering areas like technical capabilities and institutional factors). Simultaneously, we conducted an inductive analysis, openly coding the data for new, unexpected themes. Through this process, significant themes like "algorithmic bias patterns" and the "rural-urban gap" emerged.

This combined method ensured our findings were both grounded in established theory (Chap. 2) and responsive to new insights emerging from the literature. We applied the resulting hybrid codebook systematically to all studies and created thematic maps to visually show how different concepts were connected, providing a clear and reproducible path for our synthesis.

We synthesized the findings to identify:

- Prevalent computational techniques in energy governance for energy efficiency.
- Common challenges in applying computational techniques to policy analysis and coherence in developing countries.
- Successful case studies of computational techniques used for energy policy analysis.
- Methodological approaches for integrating computational methodologies into policy analysis.
- Gaps in current research indicating potential areas for future study.

These synthesis objectives guided the selection of primary and secondary data sources detailed in Sect. 3.3.

## 3.3 Data Collection Sources

Our data collection methodology ensured we study the deployment of computational tools in energy governance, specifically for improving energy efficiency. We focused on questions of policy coherence and achieving Sustainable Development Goal 7 (affordable and clean energy) in developing countries, as outlined in Table 3.6.

**Table 3.6** Primary and secondary data sources

| Source type | Specific sources | Information collected |
|---|---|---|
| Primary sources | Peer-reviewed literature on computational methods in energy policy analysis | Technical specifications of computational applications<br>Methodological approaches<br>Performance metrics |
| | Research on challenges (data accuracy, explainability, transparency) | Technical limitations<br>Ethical considerations<br>Implementation barriers |
| | Studies on policy coherence frameworks and assessment | Coherence metrics<br>Evaluation methodologies<br>Governance frameworks |
| Secondary sources | Policy briefs on SDG 7 implementation in developing countries | Implementation contexts<br>Policy priorities<br>Iinstitutional frameworks |
| | Reports from international organizations (UNDP, World Bank Group) | Case studies<br>Best practices<br>Regional analyses<br>Statistical data |
| | Conference proceedings on energy efficiency policies | Emerging approaches<br>Technological innovations<br>Policy trends |
| | Case studies of computational applications in energy policy analysis | Implementation examples<br>Success factors<br>Challenges<br>Outcomes |

For machine learning (ML) applications, we categorized techniques based on their algorithmic approaches, data requirements, and application domains. For natural language processing (NLP) applications, we analyzed techniques based on their linguistic complexity, data preprocessing requirements, and specific applications in policy document analysis.

We used triangulation across multiple data sources to ensure validity and comprehensive coverage of the research questions. We then reviewed the preliminary findings to validate the interpretations and conclusions. We grounded our results on established theories and concepts from computational methods, energy, policy development, complexity science, and sustainable development.

## 3.4 Methodological Limitations

The systematic scoping review documented in this chapter addresses empirical implementations of computational methods in energy governance, policy development, and coherence. However, the BRIDGE-E framework draws additionally upon complexity science and complexity economics, epistemic injustice theory, information ethics, energy justice, and responsible AI design principles.

We identified these theories through targeted citation searching and theoretical sampling rather than the scoping review protocol. These theories provide the necessary philosophical, ethical and conceptual foundations for our framework. This dual approach, which combines a scoping review with purposive theoretical sampling, reflects Jaakkola's (2020) framework development research. Indeed, conceptual foundations and empirical evidence serve complementary but epistemically distinct functions in theoretical contribution construction.

We acknowledge several limitations in our methodology that may affect the interpretation of results:

1. Language bias: By focusing primarily on English-language publications, we may have excluded relevant studies from non-English-speaking regions, particularly from developing countries where energy policy challenges are most acute.
2. Publication bias: Despite efforts to include gray literature, our review may underrepresent unpublished or ongoing initiatives in computational applications for energy policy.
3. Technological evolution: Given the rapid development of computational techniques, some recent innovations may not be fully represented in the published literature. This creates an inherent temporal lag between technological innovation and scholarly documentation.
4. Interdisciplinary challenges: The intersection of computational methods, policy analysis, and energy efficiency spans multiple disciplines with varying terminologies and methodological traditions. This potentially affects the comprehensiveness of our search strategy despite systematic efforts to capture cross-disciplinary contributions.
5. Scoping review boundaries: Our documented approach enables the identification of patterns and conceptual mapping appropriate to nascent fields. It introduces, however, interpretive elements in determining corpus boundaries. The documented search strategies (Tables 3.1 and 3.2), inclusion criteria (Table 3.3), quality assessment criteria (Table 3.4), and data extraction protocols (Table 3.5) provide a transparent methodology. This enables the evaluation of our analytical corpus and selection decisions. Future research advancing BRIDGE-E empirical validation should employ prospectively registered review protocols to enable comparative assessment across evolving implementation contexts.

We addressed these limitations through transparent reporting of methodology, careful interpretation of findings, and acknowledgment of areas requiring further research. The literature review provides the analytical foundation for a comprehensive examination of data science methodologies in energy policy contexts. A rigorous evaluation of peer-reviewed studies and reports meeting inclusion criteria across diverse implementations, including but not limited to developing countries, should allow reproducibility.

## 3.5 Ethical Considerations in Research Design

The ethical frameworks detailed in Chap. 2, Sect. 2.2.2 directly guided our methodological choices. Specifically:

- Fricker's (2007) epistemic justice in literature review: Our systematic review protocol included explicit searches for studies documenting marginalized perspectives and community-based knowledge systems. This targeted approach counters testimonial injustice common in technical literature. Additionally, the inclusion criteria prioritized implementations documenting contextual adaptations and local knowledge integration. The acknowledgment addresses hermeneutical injustice where marginalized communities lack conceptual resources to articulate their experiences.
- Floridi's (2013) information ethics in data collection: Our data collection strategy prioritized sources that enhance coherence, transparency, and accessibility of governance information. Our objective was to reduce informational entropy in the evidence base. The hybrid deductive-inductive thematic analysis ensures our synthesis promotes flourishing understanding rather than fragmentation.
- Dignum's (2019) responsible AI in analysis: Our analytical framework explicitly coded for ethical considerations, accountability mechanisms, and stakeholder engagement processes. This method aligns with responsibility-by-design principles that embed moral reasoning into analytical procedures.
- Jenkins et al.'s (2016) energy justice in case selection. The inclusion criteria for case studies prioritized implementations addressing distributional justice (equitable allocation of benefits and burdens), recognition justice (respect for diverse communities), and procedural justice (inclusive decision-making).

This methodology, guided by ethical principles, ensures our research not only describes existing systems but also actively works to counter the epistemic injustices, informational entropy, and equity failures they may perpetuate.

What patterns emerge when computational tools meet real-world policy complexity? Chapter 4 analyzes the evidence from the scoping review, systematically mapping how applications succeed or fail to address the governance-data-inequality nexus. This diagnosis of translational barriers provides the empirical basis for our conceptual framework developed in Part III.

## References

Arksey, H., & O'Malley, L. (2005). Scoping studies: Toward a methodological framework. *International Journal of Social Research Methodology, 8*(1), 19–32. https://doi.org/10.1080/1364557032000119616

Arthur, W. B. (1989). Competing technologies, increasing returns, and lock-in by historical events. *The Economic Journal, 99*(394), 116–131. https://doi.org/10.2307/2234208

Awogbemi, O., Von Kallon, D. V., & Kumar, K. S. (2024). Contributions of artificial intelligence and digitization in achieving clean and affordable energy. *Intelligent Systems with Applications, 22*, 200389. https://doi.org/10.1016/j.iswa.2024.200389

Dignum, V. (2019). *Responsible artificial intelligence: How to develop and use AI in a responsible way*. Springer. https://doi.org/10.1007/978-3-030-30371-6

Floridi, L. (2013). *The ethics of information*. Oxford University Press. https://doi.org/10.1093/acprof:oso/9780199641321.001.0001

Fricker, M. (2007). *Epistemic injustice: Power and the ethics of knowing*. Oxford University Press. https://doi.org/10.1093/acprof:oso/9780198237907.001.0001

Hannan, M. T., & Freeman, J. (1984). Structural inertia and organizational change. *American Sociological Review, 49*(2), 149–164. https://doi.org/10.2307/2095567

Hennink, M. M., Kaiser, B. N., & Marconi, V. C. (2017). Code saturation versus meaning saturation: How many interviews are enough? *Qualitative Health Research, 27*(4), 591–608. https://doi.org/10.1177/1049732316665344

Holland, J. H. (1995). *Hidden order: How adaptation builds complexity*. Addison-Wesley.

Jaakkola, E. (2020). Designing conceptual articles: Four approaches. *AMS Review, 10*(1), 18–26. https://doi.org/10.1007/s13162-020-00161-0

Jenkins, K., McCauley, D., Heffron, R., Stephan, H., & Rehner, R. (2016). Energy justice: A conceptual review. Energy Research & Social Science, 11, 174–182. https://doi.org/10.1016/j.erss.2015.10.004

Kauffman, S. A. (1993). *The origins of order: Self-organization and selection in evolution*. Oxford Academic. https://doi.org/10.1093/oso/9780195079517.001.0001

Levac, D., Colquhoun, H., & O'brien, K. K. (2010). Scoping studies: Advancing the methodology. *Implementation Science, 5*(1), 69. https://doi.org/10.1186/1748-5908-5-69

Munn, Z., Peters, M. D., Stern, C., Tufanaru, C., McArthur, A., & Aromataris, E. (2018). Systematic review or scoping review? Guidance for authors when choosing between a systematic or scoping review approach. *BMC Medical Research Methodology, 18*(1), 143. https://doi.org/10.1186/s12874-018-0611-x

Peters, M. D., Marnie, C., Tricco, A. C., Pollock, D., Munn, Z., Alexander, L., et al. (2020). Updated methodological guidance for the conduct of scoping reviews. *JBI Evidence Synthesis, 18*(10), 2119–2126. https://doi.org/10.11124/JBIES-20-00167

Saunders, B., Sim, J., Kingstone, T., Baker, S., Waterfield, J., Bartlam, B., et al. (2018). Saturation in qualitative research: Exploring its conceptualization and operationalization. *Quality & Quantity, 52*(4), 1893–1907. https://doi.org/10.1007/s11135-017-0574-8

Tricco, A. C., Lillie, E., Zarin, W., O'Brien, K. K., Colquhoun, H., Levac, D., et al. (2018). PRISMA extension for scoping reviews (PRISMA-ScR): Checklist and explanation. *Annals of Internal Medicine, 169*(7), 467–473. https://doi.org/10.7326/M18-0850

Vinuesa, R., Azizpour, H., Leite, I., Balaam, M., Dignum, V., Domisch, S., et al. (2020). The role of artificial intelligence in achieving the sustainable development goals. *Nature Communications, 11*(1), 233. https://doi.org/10.1038/s41467-019-14108-y

# Part II
# Empirical Analysis

In Part II, we test the complexity-ethics analytical lens through a systematic examination of computational applications, implementation contexts, and institutional initiatives. Our analysis progresses from technical capabilities to contextual manifestations to institutional validation.

Chapter 4 examines three application categories: natural language processing for semantic analysis, agent-based modeling for behavioral simulation, and network analysis for structural mapping. We note that these applications demonstrate strong technical capabilities but they reveal inherent characteristic limitations. NLP analyzes policy text but cannot model co-evolutionary discourse production. Agent-based models simulate emergence while abstracting from path-dependent institutional constraints. Network analysis maps static structures but cannot simulate dynamic fitness landscapes. We demonstrate these are not technical gaps but stable features arising from the design and evaluation of these tools.

Chapter 5 analyzes how complex adaptive systems (CAS) dynamics manifest empirically in developing countries. First, data infrastructure shows path dependence through systematic exclusion of informal knowledge and off-grid systems. Second, institutional capacity exhibits structural inertia. Our analysis reveals systemic cycles where external initiatives demonstrate feasibility then withdraw without building endogenous capability. Third, algorithmic optimization perpetuates systematic bias. Models trained on partial data generate inequitable recommendations. These three challenges operate through feedback mechanisms, each reinforcing the others.

Chapter 6 validates our framework through case studies of institutional initiatives: UNDP Accelerator Labs, World Bank ESMAP, national energy efficiency programs, and the theoretical analysis of ASEAN Centre for Energy. Despite substantial resources and explicit equity commitments, these initiatives exhibit the predicted patterns. Four requirements emerge from the systematic analysis of observed failures: operationalizing epistemic justice to counter path dependence, enabling institutional co-evolution to address structural inertia, embedding energy justice as primary optimization constraint, and respecting complexity dynamics through deliberative protocols.

Part II therefore provides an empirical validation for our diagnostic framework built on the complexity-ethics lens. This generates empirically grounded specifications for the BRIDGE-E response developed in Part III.

# Chapter 4
# Computational Applications in Energy Policy Analysis: Empirical Patterns and Translational Barriers

## 4.1 From Application Catalog to Barrier Diagnosis

Our scoping review reveals a field defined by evolving concepts and technical innovation. Current computational applications function as diagnostic instruments. They operate externally to the institutional adaptive processes they seek to inform. Our analysis started as a listing of technical capabilities in policy analysis. However, our reflection matured as we gathered empirical evidence revealing the limitations of current computational tools.

We analyzed why the design and deployment patterns of computational applications will collectively prevent them from addressing the governance-data-inequality nexus. We question whether their observed limitations represent temporary developmental gaps or stable features arising from the complex adaptive system (CAS) dynamics and ethical failures our theoretical framework explains.

We answer these questions through the complexity-ethics analytical lens developed in Chap. 2. This analytical framework foregrounds Arthur's (1989) path dependence, Hannan and Freeman's (1984) structural inertia, and the co-evolutionary dynamics of CAS (Holland, 1995; Kauffman, 1993). We synthesize these complexity science theories with the ethical principles of Fricker's (2007) epistemic injustice, Floridi's (2013) information ethics, Dignum's (2019) responsible AI, and Jenkins's (2016) energy justice.

Applying this lens provides the explanatory power to distinguish between corrigible technical shortcomings and inherent systemic pathologies. This distinction, in turn, ensures we design a complexity-aware framework with solid conceptual foundations.



R. Bouderhem, *A Computational Framework for Energy Policy Coherence*, SpringerBriefs in Energy, https://doi.org/10.1007/978-3-032-22978-6_4

## 4.2 Empirical Patterns: Three Categories of Computational Applications

The evidence converges around three core computational functions. Each demonstrates considerable technical sophistication but also distinct and characteristic barriers to practical implementation.

### *4.2.1 Natural Language Processing: Semantic Analysis of Policy Discourse*

Studies utilize computational techniques (e.g., Latent Dirichlet Allocation (LDA) and transformer-based models) to identify thematic evolution and interconnections within policy and research domains. For instance, Saheb et al. (2022) use a novel contextual topic modeling approach combining LDA and BERT to analyze the scientific literature on artificial intelligence for sustainable energy. Their analysis identifies eight dominant research topics, including sustainable buildings, climate artificial intelligence, and AI-based optimization of renewable technologies. Saheb et al. (2022) also reveal gaps and cross-topic themes within the field. They finally recommend "policymakers to design AI regulations that promote the development of sustainable algorithms and tools while mitigating the unintended consequences of unsustainable energy sources and AI solutions" (Saheb et al., 2022).

Similarly, Ashayeri and Abbasabadi (2024) use NLP and transformer models (RoBERTa) to perform sentiment analysis on social media data (X/Twitter). They assess public sentiment related to energy justice, building occupancy, and HVAC systems in New York City during the pandemic. This study demonstrates the potential of data science methodologies to diagnose urban energy justice by analyzing social media sentiment linked to building and demographic data in NYC. It reveals how computational tools can map public perception onto spatial and infrastructural inequities in real time.

At a broader scale, Wang et al. (2024) apply LDA topic modeling and bibliometric analysis to map the scholarly landscape linking renewable energy to the Sustainable Development Goals (SDGs). Their work unveils interconnected research themes such as policy uncertainty, the Environmental Kuznets Curve, and ecological footprint. They demonstrate how NLP can reveal the multifaceted intellectual structure of a field and its policy implications. These applications excel at revealing patterns within large volumes of formal and informal text, quantifying terminology, mapping topic prevalence, and measuring sentiment distributions.

All these tools function as diagnostic instruments for policy discourse. They identify synchronic patterns—discourse structure at discrete temporal points—but face inherent limitations in modeling dynamic processes. As Arthur (1989) establishes for technological systems and Levinthal (1997) extends to organizational adaptation, early choices create trajectories that become progressively entrenched through self-reinforcing mechanisms. Policy discourse operates within such dynamics. Language choices, framing strategies, and conceptual categories shape and are shaped by institutional interests and governance structures through iterative processes. NLP tools analyze discourse outputs at specific moments but cannot model the co-evolutionary processes through which those outputs continuously reconstitute themselves within adaptive governance systems.

For instance, the tool deployed by Ashayeri and Abbasabadi (2024) operates as an external diagnostic tool. It analyzes sentiment about the system, but there is no integration into the governance processes. It maps the symptom (public sentiment on inequality) but does not, and cannot by design, intervene in the complex adaptive system (CAS) of urban governance, policy, or resource allocation to address the root causes. Thus, it serves as a prime example of a sophisticated tool that, while revealing, remains confined to the periphery of the governance-data-inequality nexus it helps to illuminate. It identifies patterns within discourse rather than engaging with the institutional dynamics producing that discourse.

### *4.2.2 Agent-Based Modeling: Behavioral Simulation in Abstracted Environments*

A second approach applies machine learning (ML) and agent-based modeling for predictive and exploratory analysis. Wang et al. (2023) developed a CNN-LSTM model that integrates numerical time-series data with multimodal text data (policy documents, news, forum comments). Their objective is to improve the accuracy of forecasting China's national power demand, which is a critical input for decarbonization planning.

As shown by Perera and Kamalaruban (2021), Reinforcement Learning (RL) has significant, yet underutilized, potential to manage the growing complexity of modern energy systems. This is due to renewable integration, sector coupling, and cyber-physical interactions. Their review conducts a top-down analysis to map the state of RL applications, revealing a performance gap compared to advanced RL methods used in other fields. Based on this study, the transition to complex, decarbonized energy systems presents a grand challenge for control optimization (Perera & Kamalaruban, 2021). These authors conclude that RL is a powerful candidate to address it, but its application in the energy domain must mature significantly to keep pace with both the growing complexity of energy systems and the rapid advances in RL itself.

These applications serve as exploratory simulators, which generate projections within computationally defined environments. They successfully model emergence, specifically how micro-level interactions produce macro-level patterns that cannot be deduced from component behaviors alone (Holland, 1998). However, they misrepresent the context of governance. Why? Because they simulate agent behaviors under idealized assumptions, projecting outcomes as if actors (e.g., institutions, corporations, local communities) operate free from constraint. There is indeed a disconnection between current computational tools and real policy implementation. These models ignore historical constraints that shape policy.

Policy implementation traditionally operates under path dependence (Arthur, 1989), where past infrastructure and policy choices lock in future options. Structural inertia (Hannan & Freeman, 1984) is another characteristic of policy implementation as organizational routines powerfully resist change. Consequently, these simulators answer "what could happen" in an abstract system, while effective governance requires understanding "what will happen" within the actual, constrained institutional environment.

### *4.2.3 Network Analysis and Algorithmic Inference: Structural Diagnostics on Static Landscapes*

This category of applications focuses on mapping policy systems. Network analysis, for instance, visually charts connections within complex nexuses like water-energy-food governance (Weitz et al., 2017). Other algorithmic approaches infer patterns from development data. A sophisticated example is the Policy Priority Inference (PPI) framework, which uses agent-based modeling to infer the historical budgetary priorities that generated observed changes in Sustainable Development Goal indicators (Guerrero & Castañeda, 2020).

These applications are powerful diagnostic instruments. They generate authoritative, data-driven snapshots of structural relationships (conflicts, synergies, central nodes) within a policy system at a point in time.

The PPI framework illustrates this clearly. Although explicitly built on complexity principles, including agent learning within the simulation, its design is that of a closed analytical model. It does not enable meta-learning or co-evolution. Moreover, its institutional parameters and structural network are fixed inputs, not dynamically shaped by stakeholder deliberation or real-world policy feedback. Thus, the model's internal dynamics remain self-contained. They do not facilitate an adaptive dialogue with the real-world processes of governance such as political negotiation, institutional adaptation, and power realignment.

This self-containment is a fundamental flaw because it makes the tools blind to the co-evolutionary dynamics that define real governance. Kauffman's (1993) fitness landscape theory, extended by Levinthal (1997), establishes that strategic fitness is dependent on a continuously reconfigured environment of institutions, coalitions, and technologies. A network diagram or a model's output is a frozen moment in this adaptive process. It can diagnose a structural condition but cannot simulate, or integrate into, the political dynamics through which that structure itself evolves.

Therefore, such tools remain analytically external to the adaptive governance processes they analyze. This directly aligns with Weitz et al.'s (2017) finding that effective nexus governance requires continuous adaptation, a requirement that static structural maps and even the most advanced diagnostic simulations are unable to fulfill.

## 4.3 Translational Barriers: CAS Dynamics in Practice

The applications' limitations are direct consequences of CAS dynamics. They operate at the level of development and deployment. These dynamics manifest in three interconnected ways.

### *4.3.1 Path Dependence in Data Infrastructure*

Arthur's (1989) path dependence theory explains how early choices in system development become locked-in through self-reinforcing mechanisms. This dynamic manifests clearly in computational tools' epistemological confinement to formal, structured, and institutionally legible data. NLP applications require digitized text; machine learning (ML) models rely on quantitative time-series; network analysis needs codified policy attributes. Current computational tools optimize for formal data while structurally excluding tacit, experiential, and community-held knowledge.

The consequences manifest empirically. Salhi et al. (2025) are using AI "to create sustainable and carbon-free urban environments." They achieve near-perfect predictive performance by relying on consolidated, quantitative public datasets (e.g., UCI ML Repository) representative of formal, documented systems like building energy efficiency and grid-connected household consumption. This technical success is epistemologically contingent because the models limit the analytical scope to documented realities within dominant data schemas. Informal settlements, off-grid systems, decentralized microgrids, and community-managed energy solutions lack such formal data representation. As such, they remain outside the models' operational ontology.

This situation exemplifies Fricker's (2007) hermeneutical injustice. These computational applications systematically exclude realities that lack formal categorical representation within dominant knowledge systems. This exclusion prevents these realities from entering collective understanding. Consequently, communities without formal data infrastructure are not merely underserved; they become algorithmically invisible. Their energy needs remain unrepresented in the computational optimization processes that increasingly drive infrastructure investment decisions.

From Floridi's (2013) information ethics perspective, this represents informational entropy within governance systems. Computational methodologies generate vast data about formal systems. However, they make it harder to see or account for marginalized realities. This active exclusion directly violates the constitutive dimensions of energy justice (Jenkins et al., 2016): distributional equity, procedural fairness, and recognition of diverse communities. Arthur's (1989) path dependence in data infrastructure breaches all three: resources flow toward populations already represented in data, decision-making excludes those lacking a computational voice, and system architectures fail to recognize diverse community realities.

### 4.3.2 Structural Inertia in Evaluation Metrics

Hannan and Freeman's (1984) structural inertia theory explains how organizations resist transformation due to reliability and accountability imperatives. This dynamic is evident in the evaluation frameworks that shape and validate computational tools. Success is defined by technical metrics, such as predictive accuracy ($R^2$, MAE, RMSE), classification performance (F1-score, AUC-ROC), computational efficiency, and algorithmic sophistication. These metrics are fundamentally different from, and often work against, the core imperatives of governance success. Those imperatives are policy legitimacy, distributional equity, institutional adaptability, democratic accountability, and procedural justice.

Advancements in specific computational techniques represent genuine technical progress. Notable examples include novel forecasting architectures (Wang et al., 2023), performance improvements in reinforcement learning for energy system control (Perera & Kamalaruban, 2021), and new explainable AI techniques for building energy classifiers (Galli et al., 2022). However, the abovementioned technical metrics defining success in these domains cannot measure governance outcomes. They cannot determine whether a forecast enables a just transition, whether a performance improvement benefits marginalized communities equitably, or whether an explanation fosters democratic accountability rather than just technical transparency.

This situation creates a structural mismatch. Although computational tools evolve by optimizing for stable, quantifiable technical performance, governance requires navigating dynamic landscapes where the definition of "optimal" shifts through political negotiation. Consequently, the field makes tangible technical progress on measurable dimensions. However, these tools remain peripheral to the adaptive challenges of governance.

Rogge et al. (2017) demonstrate that effective energy transition governance requires attention to systemic characteristics across policy mixes, not merely individual instrument performance. The structural inertia in technical metrics neglects these systemic dimensions. Models may achieve exceptional predictive accuracy but remain silent on whether policy instruments contradict each other, are viewed as legitimate by all stakeholders, or align with strategic objectives. Technical optimization evaluates component performance, while governance challenges demand systemic coherence.

This represents a failure to integrate moral reasoning within technical success criteria, a direct violation of responsibility by design (Dignum, 2019). In the prevailing evaluation structure, we apprehend accuracy and efficiency as intrinsic goods. Their value, however, is not intrinsic; it is instrumental. It depends entirely on whose interests they serve and which governance objectives they advance.

### *4.3.3 Co-Evolutionary Mismatches between Computational and Institutional Timelines*

Co-evolution (Kauffman, 1993) on dynamic fitness landscapes, extended to organizational adaptation (Levinthal, 1997), explains how components in complex systems must continuously adjust to each other's changes. Computational tools reveal co-evolutionary characteristics with governance systems. Institutions evolve through political negotiation, bureaucratic learning, and power realignments. Computational systems, however, operate on standardized technical cycles optimized for efficiency and scalability.

This temporal disconnect shapes the design of computational outputs, prioritizing bureaucratic assimilation over institutional reform. Tools generate data (e.g., reports, statistics) that conform to established reporting formats and sectoral review cycles. The result is a form of technocratic reinforcement. Data generated serve to legitimize and harden the existing departmental boundaries, jurisdictional mandates, and power distributions that constitute fragmentation.

The consequence is a gap between diagnostic and policy prescription. Computational tools excel at diagnosing fragmentation. As shown, they map policy conflicts and quantify inconsistencies with precision. However, the solutions they generate are invariably procedural. They simply recommend enhanced coordination or collaboration. Such incremental adjustments work within existing rules and power structures. Therefore, they avoid any confrontation with the path-dependent institutional architectures that are the root cause of the fragmentation.

This limitation aligns precisely with integration failures identified by governance theory. Weitz et al. (2017) reveal three persistent gaps in water-energy-food nexus governance: "1) conditions for cross-sector coordination and collaboration; 2) dynamics beyond cross-sector interactions; and 3) political and cognitive factors as determinants of change." Computational tools, despite their analytical sophistication, are by design constrained from addressing this full spectrum. They can technically model coordination (gap 1) but cannot engage the broader institutional dynamics (gap 2) or navigate the power structures and belief systems (gap 3) that ultimately determine policy outcomes.

These applications enable "policy patching" (Kern et al., 2017), where tools are layered onto static governance structures. This process cannot produce the "policy packaging" (Kern et al., 2017) that integrates instruments with a strategic vision, a necessary condition for policy coherence. As Rogge et al. (2017) demonstrate, effective transition governance demands precisely this deep strategic alignment across objectives, targets, and execution, an outcome unattainable through incremental technical additions alone.

Consequently, computational tools evolve into the conservative instruments that path dependence (Arthur, 1989) and structural inertia (Hannan & Freeman, 1984) predict. They allow organizations to adopt the veneer of analytical innovation without disrupting the bureaucratic routines, jurisdictional silos, and power structures at the root of policy incoherence. Their design, which prioritizes assimilation, ultimately sustains existing governance frameworks through a performance of change rather than substantive reform.

## 4.4 Synthesis: Translational Barriers as Stable System Features

The three categories of applications, bound by three manifestations of CAS dynamics, constitute a stable, self-reinforcing pattern. This deepens the path dependencies and structural inertia, making tools progressively more useful to—and firmly locked into—existing institutional logics. This represents translational barriers not as passive gaps but as active equilibrium states maintained by self-reinforcing dynamics.

These applications expertly diagnose governance challenges while remaining structurally incapable of facilitating the institutional co-evolution, power renegotiation, and adaptive processes required to resolve those challenges. This dichotomy between diagnosis and resolution echoes a fundamental insight from governance theory. As demonstrated in Sect. 4.3.3, this aligns with the empirical analysis of nexus governance by Weitz et al. (2017), where computational tools only address technical coordination problems while remaining excluded from the political and cognitive dynamics essential for systemic change. Consequently, current computational applications can only diagnose a problem. Their external, noninteractive design precludes the continuous feedback and deliberative integration essential for transformation.

Evaluated through complexity science, current applications address selected CAS dimensions like emergence simulation, static structural mapping, and synchronic pattern recognition. Notably, they systematically omit dynamics that explain governance failure: path dependence (Arthur, 1989), structural inertia (Hannan & Freeman, 1984), and co-evolutionary transformation across dynamic landscapes (Kauffman, 1993; Levinthal, 1997).

From an ethical perspective, they not only fail to remedy but actively perpetuate epistemic injustice (Fricker, 2007), generate informational entropy (Floridi, 2013), violate energy justice principles (Jenkins et al., 2016), and fail to integrate moral reasoning within computational architecture (Dignum, 2019). From the policy mix perspective, they evaluate instrument performance while neglecting systemic characteristics (consistency, coherence, credibility) that define effective transition governance (Rogge et al., 2017).

## 4.5 Implications for Architectural Specifications

The problem is not a matter of technical capability, inadequate data availability, or limited algorithmic sophistication. The problem is the manifestation of CAS dynamics (path dependence, structural inertia, co-evolutionary mismatches) combined with systematic ethical failures at the architectural level. Applications locked into external diagnostics, formal epistemologies, misaligned success metrics, and institutional assimilation cannot address the governance-data-inequality nexus or facilitate the coherent, adaptive, and just governance of complex energy transitions.

This systemic pattern establishes specific design specifications for any framework seeking to address these translational barriers. The imperative is not for incremental refinement like developing accurate predictors, efficient algorithms, and comprehensive visualizations. The imperative is for meticulous responses to each manifestation of CAS dynamics and ethical failure.

The needed framework must:

1. Counter path dependence (Arthur, 1989) in data infrastructure through tacit, experiential, and community-held knowledge. This means expanding hermeneutical resources (Fricker, 2007) by designing computational spaces where marginalized knowledge systems define algorithmic categories, rather than being forced into predefined schemas. It also requires operationalizing energy justice principles (Jenkins et al., 2016) into constitutive design criteria, not post-hoc evaluations.
2. Overcome structural inertia (Hannan & Freeman, 1984) by integrating governance success metrics into computational optimization. This means equity, legitimacy, adaptability, and the systemic characteristics of a coherent policy mix (Rogge et al., 2017) must become intrinsic reward signals, not external checks. This embodies Dignum's (2019) responsibility-by-design, where moral reasoning directs the search for solutions. Consequently, the system optimizes for outcomes that build trust, ensure strategic alignment, and enable transformation beyond incremental adjustment.
3. Resolve co-evolutionary mismatches (Kauffman, 1993; Levinthal, 1997) by designing systems for institutional co-evolution. The computational design must enable "policy packaging" rather than facilitating "policy patching" (Kern et al., 2017). It must therefore support the development of coherent "policy strategies" and sustain the ongoing "policy processes" through which coherence is achieved (Rogge et al., 2017).

These are not technical specifications for better algorithms. They are imperatives, derived directly from our empirical analysis of CAS dynamics and ethical failure. Consequently, they mandate a system whose core logic is adaptive, which treats complexity as a design parameter, uses computation to support human judgment, and institutionalizes energy justice (Jenkins et al., 2016) as a constitutive constraint.

The translational barriers documented here create a clear mandate. They define the design challenge for the BRIDGE-E framework. New computational approaches should break the stable, pathological equilibrium of current tools.

Chapter 5 details the constraints of data, capacity, and ethics that shape it. With this complete diagnosis (Chaps. 4 and 5), we can design a framework aimed at transformation, not perpetuation, of the status quo.

## References

Arthur, W. B. (1989). Competing technologies, increasing returns, and lock-in by historical events. *The Economic Journal, 99*(394), 116–131. https://doi.org/10.2307/2234208

Ashayeri, M., & Abbasabadi, N. (2024). Unraveling energy justice in NYC urban buildings through social media sentiment analysis and transformer deep learning. *Energy and Buildings, 306*, 113914. https://doi.org/10.1016/j.enbuild.2024.113914

Dignum, V. (2019). *Responsible artificial intelligence: How to develop and use AI in a responsible way*. Springer. https://doi.org/10.1007/978-3-030-30371-6

Floridi, L. (2013). *The ethics of information*. Oxford University Press. https://doi.org/10.1093/acprof:oso/9780199641321.001.0001

Fricker, M. (2007). *Epistemic injustice: Power and the ethics of knowing*. Oxford University Press. https://doi.org/10.1093/acprof:oso/9780198237907.001.0001

Galli, A., Piscitelli, M. S., Moscato, V., & Capozzoli, A. (2022). Bridging the gap between complexity and interpretability of a data analytics-based process for benchmarking energy performance of buildings. *Expert Systems with Applications, 206*, 117649. https://doi.org/10.1016/j.eswa.2022.117649

Guerrero, O. A., & Castañeda, G. (2020). Policy priority inference: A computational framework to analyze the allocation of resources for the sustainable development goals. *Data & Policy, 2*, e17. https://doi.org/10.1017/dap.2020.18

Hannan, M. T., & Freeman, J. (1984). Structural inertia and organizational change. *American Sociological Review, 49*(2), 149–164. https://doi.org/10.2307/2095567

Holland, J. H. (1995). *Hidden order: How adaptation builds complexity*. Addison-Wesley.

Holland, J. H. (1998). *Emergence from chaos to order*. https://doi.org/10.1093/oso/9780198504092.001.0001. Online edn, 31 Oct. 2023, Accessed 20 Dec. 2025.

Jenkins, K., McCauley, D., Heffron, R., Stephan, H., & Rehner, R. (2016). Energy justice: A conceptual review. *Energy Research & Social Science, 11*, 174–182. https://doi.org/10.1016/j.erss.2015.10.004

Kauffman, S. A. (1993). *The origins of order: Self-organization and selection in evolution*. Oxford Academic. https://doi.org/10.1093/oso/9780195079517.001.0001

Kern, F., Kivimaa, P., & Martiskainen, M. (2017). Policy packaging or policy patching? The development of complex energy efficiency policy mixes. *Energy Research & Social Science, 23*, 11–25. https://doi.org/10.1016/j.erss.2016.11.002

Levinthal, D. A. (1997). Adaptation on rugged landscapes. *Management Science, 43*(7), 934–950. https://doi.org/10.1287/mnsc.43.7.934

Perera, A. T. D., & Kamalaruban, P. (2021). Applications of reinforcement learning in energy systems. *Renewable and Sustainable Energy Reviews, 137*, 110618. https://doi.org/10.1016/j.rser.2020.110618

Rogge, K. S., Kern, F., & Howlett, M. (2017). Conceptual and empirical advances in analysing policy mixes for energy transitions. *Energy Research & Social Science, 33*, 1–10. https://doi.org/10.1016/j.erss.2017.09.025

Saheb, T., Dehghani, M., & Saheb, T. (2022). Artificial intelligence for sustainable energy: A contextual topic modeling and content analysis. *Sustainable Computing: Informatics and Systems, 35*, 100699. https://doi.org/10.1016/j.suscom.2022.100699

Salhi, A., Algarni, F., Alshamrani, R., et al. (2025). Leveraging artificial intelligence to enable sustainable urban development through the creation of smart and environmentally friendly carbon-free cities. *Scientific Reports, 15*, 35791. https://doi.org/10.1038/s41598-025-16801-z

Wang, D., Gan, J., Mao, J., Chen, F., & Yu, L. (2023). Forecasting power demand in China with a CNN-LSTM model including multimodal information. *Energy, 263*(Part E), 126012. https://doi.org/10.1016/j.energy.2022.126012

Wang, Q., Huang, R., & Li, R. (2024). Renewable energy and sustainable development goals: Insights from latent dirichlet allocation thematic and bibliometric analysis. *Sustainable Development, 32*(6), 6339–6359. https://doi.org/10.1002/sd.3027

Weitz, N., Strambo, C., Kemp-Benedict, E., & Nilsson, M. (2017). Closing the governance gaps in the water-energy-food nexus: Insights from integrative governance. *Global Environmental Change, 45*, 165–173. https://doi.org/10.1016/j.gloenvcha.2017.06.006

# Chapter 5
# Empirical Manifestations of Translational Barriers in Developing Contexts

## 5.1 Introduction: Empirical Evidence of Theoretical Dynamics

Chapter 4 established that computational applications are constrained by three core CAS (Holland, 1995, 1998) dynamics: Arthur's (1989) path dependence in data infrastructure, Hannan and Freeman's (1984) structural inertia in institutional evaluation, and co-evolutionary (Kauffman, 1993; Levinthal, 1997) mismatches between technical and governance timescales. These dynamics enable Fricker's (2007) epistemic injustice, Floridi's (2013) informational entropy, and violations of Jenkins et al.'s (2016) energy justice diagnosed in Chap. 2.

This chapter examines how this theoretical triad manifests empirically in developing contexts. The evidence reveals three interconnected implementation challenges that give concrete, operational form to the governance-data-inequality nexus. Analyzing them is essential for deriving specifications for any framework that must treat these dynamics as first-order design constraints, not as secondary obstacles.

## 5.2 Arthur's Path Dependence in Data Infrastructure

**The Misdiagnosis**

Conventional narratives frame the problem as one of "data scarcity"—a quantitative deficit solvable by more collection. This approach misdiagnoses the core issue, which is a structural incompleteness arising from path dependence (Arthur, 1989). Data architectures are not neutral; these applications optimize bureaucratic legibility and organizational accountability, not holistic or equitable policy assessment.

The path dependence begins with early, formative choices in data infrastructure design. The focus is on what to measure, how to categorize, which populations and



R. Bouderhem, *A Computational Framework for Energy Policy Coherence*, SpringerBriefs in Energy, https://doi.org/10.1007/978-3-032-22978-6_5

activities to formalize. As such, we cannot easily reverse these choices. Lock-in emerges through the self-reinforcing mechanisms Arthur (1989) identified:

1. Increasing returns: each new application or database is built for compatibility with existing systems. This continuous investment makes established data schemas progressively more valuable and entrenched.
2. Switching costs: personnel develop expertise in the existing ontologies. Changing the fundamental categories requires retraining, system overhauls, and procedural redesign—costs that organizations are structured to avoid.
3. Network effects: the value of a data schema increases as more agencies adopt it for interoperability. This creates powerful incentives for new systems to conform to old standards, making unilateral innovation by a single department nearly impossible.

Together, these mechanisms ensure that initial design choices, often made for administrative convenience or historical accident, congeal into a rigid infrastructure. This infrastructure, in turn, determines what counts as "data" and systematically excludes realities that do not fit its formal categories.

**Empirical Manifestation**

Data exists, but in fragmented and nonstandardized silos. This is a direct reflection of institutional fragmentation. Debnath and Mourshed (2018), in their comprehensive review of energy planning models (EPMs), identify a systematic inadequacy for developing-country contexts rooted in deficiencies of data quality, availability, and adequacy. Their analysis documents how such models fail to account for socio-economic nuances critical to development settings.

This problem is entrenched in data infrastructure. Energy ministries, statistical offices, and utilities often maintain incompatible databases built on divergent conceptual frameworks. Consequently, data collection schemas systematically capture formal, grid-connected, and revenue-generating activities, while excluding informal economies, off-grid energy use, and community-level practices. A parallel challenge exists in broader development metrics. Nilashi et al. (2023) document that for Sustainable Development Goal (SDG) measurement, pervasive data incompleteness degrades quality and hinders accurate performance tracking.

Read together, these analyses reveal a deeper, systemic issue. Data incompleteness is often not random but structured. The architecture of data systems encodes a specific ontology, one that legitimizes energy systems as grid-connected infrastructure serving formal economic actors. Realities that deviate from this schema are systematically excluded from categorical representation, lacking conceptual legitimacy within official taxonomies. This is the case for instance for off-grid renewable microgrids, community-managed cooperatives, informal sector energy use, and traditional biomass consumption. This ontological bias has direct modeling consequences. As Debnath and Mourshed (2018) highlight, a critical omission is the absence of deliberation on suppressed energy demand resulting from a lack of access and supply.

The calibration of energy planning models on the consumption patterns of already-served populations has direct consequences. It renders the unmet demand of unserved populations invisible in policy and investment forecasts.

**How Arthur's Path Dependence Operates**
Early choices in data infrastructure, optimizing for bureaucratic accountability and fiscal governance, become progressively entrenched through the self-reinforcing mechanisms described above. The resulting inertia systematically excludes nonconforming realities, concretizing informational entropy (Floridi, 2013). Data ecosystems do generate high volumes of structured information, but at the same time they degrade the capacity to represent marginalized realities. An immediate consequence is hermeneutical injustice at scale (Fricker, 2007). The erasure of lived experiences and energy access challenges of marginalized populations is epistemic. The conceptual frameworks that define "energy policy problems" and measure "energy system states" exclude these populations.

**Impact on AI Systems**
AI models trained on these path-dependent datasets suffer from constitutive sampling bias. They become expert models of formal, grid-connected, urban sectors while remaining ontologically blind to Indigenous populations and marginalized realities. This epistemic failure originates in the foundational stage that Liang et al. (2022) identify as critical for trustworthy AI—data design. When the design of training corpora results from sources that systematically exclude informal settlements, off-grid systems, and community-managed solutions, no subsequent algorithmic sophistication can rectify the resulting ontological blindness. The model learns a partial ontology as its complete reality.

**Required Response**
Addressing Arthur's path dependence requires designing for epistemic inclusion. We need to develop computational systems capable of ingesting, respecting, and reasoning with heterogeneous data forms. Therefore, we need to account for qualitative narratives from community consultations, nonstandardized local surveys, participatory mapping exercises, and traditional knowledge documentation. Epistemic justice (Fricker, 2007) and responsibility-by-design (Dignum, 2019) must be constitutive principles rather than post-hoc corrections.

## 5.3 Hannan and Freeman's Structural Inertia in Institutional Capacity Development

**The Pattern**
Externally funded technology-for-development initiatives exhibit a systematic failure pattern. They demonstrate technical feasibility, produce deliverables and reports, and conclude. They routinely fail to catalyze a transition into sustained, nationally owned institutional capabilities. This outcome manifests Hannan and Freeman's

(1984) structural inertia. Government ministries and agencies, for instance, possess powerful, embedded defenses against transformational change, prioritizing reliability, accountability, and the reproduction of existing routines.

**Empirical Manifestation of the Failure Pattern**
Unwin's (2017) analysis of Information and Communication Technology for Development (ICT4D) initiatives documents this pattern. Projects designed to reduce poverty consistently fail to achieve scale and prove unsustainable upon the cessation of donor funding. Unwin diagnoses a structural cause. The external conception and governance of these initiatives undermine endogenous institutional development.

This failure pattern replicates in data-driven governance. Kim et al.'s (2023) study of 133 data-based policy innovation labs (PILs) provides corroboration. Their analysis finds that such labs enhance analytical and operational policy capacity. However, they consistently fail to build political capacity, the authority and legitimacy necessary for institutionalization (Kim et al., 2023). This deficit suggests that analytical gains remain fragile, externally dependent, and vulnerable to dissipation when external support ends.

**The Endogenous Condition: Institutional Barriers as the Root Cause**
The ultimate barrier to sustainable adoption is not the external project design alone but the structural inertia of the target institutions themselves. Sharma et al.'s (2022) empirical investigation of AI implementation challenges within India's public manufacturing sector identifies the endogenous condition. Deep-seated institutional barriers include "poor data quality, managers' lack of understanding of cognitive technologies, data privacy" and prohibitive costs (Sharma et al., 2022). These barriers reflect institutional resistance to changes that threaten established routines, expertise, and accountability structures.

Externally driven initiatives focused on demonstrating technical solutions are typically ill-equipped to dismantle these core institutional defenses. This creates the observed mismatch. A project may successfully produce a report on AI's potential (an external deliverable) while failing to transform the internal conditions (Sharma et al.'s barriers) that block its sustained use.

**How Hannan and Freeman's Structural Inertia Operates**
Organizations resist the integration of transformative external innovations through the mechanisms Hannan and Freeman (1984) identify:

1. Reliability imperative: bureaucracies survive by prioritizing consistent, reproducible processes. Complex external AI systems introduce technical and operational unpredictability. This constitutes a dependence on technologies the organization cannot maintain, expertise it cannot replace, and decision processes it cannot fully control. The rational organizational response is rejection or superficial, nonintegrated adoption.
2. Accountability imperative: public sector organizations are accountable for failures. Adopting sophisticated AI systems that staff lack the capacity to understand, audit, or defend in policy or public forums creates accountability risks.

The safer course is minimal, symbolic adoption. Typically, this would be using AI for external reporting or pilots while retaining traditional, defensible decision-making workflows internally.

3. Sunk costs in existing capabilities: organizations have substantial investments in incumbent analytical methods, personnel trained in established practices, and workflows designed around legacy tools. A full transformation to AI-centric processes demands the costly abandonment of these sunk investments. Structural inertia powerfully favors the addition of AI as an optional, supplementary tool over its integration as a transformative core capability.

**Co-evolutionary Mismatch and Path-Dependent Filtering**

The clash between external project logic and internal institutional logic creates a co-evolutionary mismatch (Kauffman, 1993; Levinthal, 1997). Casula's (2022) analysis of transformative innovation policy in the European Union demonstrates the mechanism. National institutional contexts act as path-dependent filters, diluting and reshaping external policy mandates to align with domestic traditions and structures.

Applied to technology projects, this means the rapid, output-oriented lifecycle of an external initiative (focused on deliverables, publications, and technical demonstrations) is structurally incompatible with the slow, path-dependent timeline of genuine institutional learning and adaptation. Lock-in and path dependence (Arthur, 1989) emerge through early technical choices around architecture and data. The organization's reliability and accountability imperatives (its structural inertia) actively resist the disruptive integration process. The result is not transformation but technological dependency. External technical expertise becomes a permanent crutch rather than a transitional bridge to endogenous capability.

**Self-Reinforcing Cycle of Dependency**

This dynamic operates through a mutually reinforcing cycle:

1. Capacity gap framing: a perceived "capacity gap" is framed in technical terms. This justifies an externally designed solution that overlooks underlying governance and institutional deficits.
2. External delivery on external terms: the solution is delivered according to external priorities and methodologies. Gelan et al.'s (2024) analysis of AI governance decolonization in Sub-Saharan Africa documents the prevalent stagnation. Among 10 evaluated countries, 80% remain categorized as "decolonization-aware" without demonstrating substantive progress in building endogenous governance frameworks, with only one nation (Rwanda) noted as notably responsive.
3. Dependency reinforcement: external experts operating sophisticated systems inadvertently reinforce the perception of local incapacity. Local institutions are positioned as consumers, not co-creators, and their existing contextual knowledge is devalued.
4. Dissipation: the project concludes. The infrastructure, lacking local stewardship and adaptation, becomes inoperable. Expertise departs. Any nascent capacity dissipates.

5. Cycle continuation: the demonstrated "failure" of the local institution to sustain the intervention is used to justify the next cycle of external delivery, further precluding the possibility of endogenous capability development.

**Consequence: The Foreign Implant**

The ultimate consequence is precisely what institutional theory predicts. AI remains a foreign implant. It may produce outputs for external reporting but fails to integrate into the core routines of governance, which are deliberation, budget negotiation, and political contestation. It cannot address the translational barriers identified by Sharma et al. (2022) because it never becomes part of the institution's own adaptive repertoire.

**Required Response: Designing for Co-evolution**

Addressing this structural inertia requires moving beyond technical solutionism (Lindgren and Dignum, 2023) to design for institutional co-evolution. The foundational principle must be capacity co-production. This necessitates:

- Hybrid development teams: integrating external technical expertise with permanent institutional staff from inception.
- Open and modifiable architectures: employing open-source systems that enable local understanding, modification, and ownership.
- Binding funding models: structuring external capital around sustained institutional partnerships and milestones of endogenous capability growth, not short-term deliverables.

This approach applies Dignum's (2019) responsibility-by-design at the systemic level, making epistemic justice (Fricker, 2007) and institutional empowerment constitutive architectural principles. It designs systems that require, rather than merely permit, the co-evolution of technology and institution, directly countering the path-dependent inertia that perpetuates dependency.

## 5.4 Convergent Manifestation: Algorithmic Bias as Systematic Ethical Failure

**The Nature of the Challenge**

Algorithmic bias in this context is neither a statistical artifact nor an abstract ethical concern. It is the operational outcome of convergent systemic failures. First, Arthur's (1989) path dependence supplies biased training corpora from historical data infrastructure. Second, Hannan and Freeman's (1984) structural inertia ensures institutional objectives prioritize external, technical metrics over equitable governance. This convergence generates inequitable policy through computational optimization.

**Empirical Manifestation**

Consider an AI model for national grid investment planning trained on historical connection data from path-dependent infrastructure (capturing only formal, grid-connected, revenue-generating installations) and optimized for conventional economic metrics (cost-per-connection, predicted revenue-to-cost ratios). This model will systematically recommend grid extension to commercial districts and peri-urban middle-class areas over decentralized renewable solutions for remote, low-revenue villages.

This outcome is not an error. It is the model correctly executing its design, optimizing its given function on the biased distribution it was trained on.

**Amplification by Documented Human Overreliance**

Human behavior amplifies the problematic outputs of such systems. Klingbeil et al.'s (2024) study demonstrates that the mere knowledge of advice being AI-generated causes overreliance. Participants "follow AI advice even when it contradicts available contextual information as well as their own assessment," leading to inefficient outcomes. Wellstead et al. (2024) also confirm that the use of AI tools should not be a "substitute for human analysis but rather a complementary tool." In policy translation, this means planners may accept recommendations that systematically disadvantage populations absent from the model's logic.

Furthermore, the deployment mechanism itself affects legitimacy. Starke and Lünich's (2020) investigation on political decision-making in the EU finds that while hybrid decision-making (politicians + AI) maintains legitimacy, algorithmic systems as sole decision-makers are perceived as illegitimate.

**Systematic Violations of Ethical Frameworks**

This convergent failure manifests as violations across four ethical frameworks:

1. Fricker's (2007) epistemic injustice: the system enacts testimonial injustice (marginalized populations' energy needs are absent from training data) and hermeneutical injustice (their realities lack the conceptual representation to be counted in optimization criteria).
2. Floridi's (2013) information ethics: the system increases informational entropy. It generates voluminous, high-precision data and recommendations that actively degrade the information ecosystem's capacity to represent marginalized energy realities.
3. Jenkins et al.'s (2016) energy justice: the model's optimization violates all three tenets: distributional justice (allocating resources away from the underserved), recognition justice (failing to acknowledge diverse community energy realities), and procedural justice (excluding affected populations from the setting of algorithmic priorities).
4. Dignum's (2019) responsible AI: the system fails this core principle by not embedding moral reasoning. It treats technical accuracy and efficiency as intrinsic goods, rather than instrumental capacities whose value is contingent on whose interests they serve.

**Impact on Policy and the Limits of "Explainability"**
Consequently, the AI system functions as a technocratic amplifier of existing inequality. Its "evidence-based" outputs lend an objective veneer (statistical rigor, computational precision) to decisions that systematically disadvantage marginalized groups.

In such contested policy domains, technical transparency alone is inadequate. De Bruijn et al. (2022) conclude that for "wicked" policy problems, characterized by ambiguous facts and contested normative criteria, explainable decisions are often not perceived as objective. As de Bruijn et al. (2022) argue, technical transparency about how a model computes an output does not address why the objective function and training data embed historical biases, failing to build public trust or legitimacy in contested domains.

**Required Response: Architecting for Justice-by-Design**
Addressing this requires moving from ethics-by-checklist to justice-by-design. Equity metrics must be first-order optimization constraints, not post-hoc evaluations. It necessitates:

- Contestable objective functions: architectures that allow affected communities to audit and challenge the normative priorities (e.g., cost vs. access) baked into algorithmic objectives.
- Transparent and compliant architectures: systems built to adhere to emerging regulatory mandates for AI, employing technical frameworks for explainability and accountability as discussed in works like Papadakis et al. (2024).
- Deliberative scaffolding: institutional designs that position AI as a tool to support human deliberation and political judgment. We need to inform the democratic process rather than seeking to circumvent or replace it.

## 5.5 Synthesis: The Interdependent System

These three challenges are not isolated. They form an interdependent system that materially enacts the governance-data-inequality nexus. Their systemic nature becomes clear through three reinforcing feedback mechanisms.

**Feedback 1: Data Infrastructure → Algorithmic Bias**
Arthur's (1989) path dependence in data infrastructure produces epistemically biased training corpora. Collection schemas optimized for bureaucratic legibility systematically exclude informal economies, off-grid systems, and community-managed solutions. AI models trained on these corpora become expert systems for this partial reality. When deployed, they generate biased recommendations with structural inevitability. They have learned, with high precision, to optimize for populations represented in the data.

**Feedback 2: Structural Inertia → Dependency → Data**

Hannan and Freeman's (1984) structural inertia perpetuates technological dependency through cycles where external project delivery precludes institutional ownership. Without endogenous computational capacity, institutions cannot challenge the ontological assumptions rooted in data schemas, advocate for including marginalized realities, or modify collection priorities to serve equity objectives. Consequently, the path-dependent data infrastructure persists precisely because the institutions that could transform it lack the capacity to articulate alternative architectures.

**Feedback 3: Algorithmic Bias → Resource Allocation → Governance**

Biased algorithms generate resource allocation recommendations (e.g., for grid investments, subsidies, technical assistance) that systematically favor formally represented populations. These outputs, delivered with the veneer of computational objectivity, lend technocratic legitimacy to decisions that reinforce existing inequalities.

As Klingbeil et al. (2024) demonstrate, the "mere knowledge that advice" is AI-generated causes overreliance. Policymakers, lacking the endogenous capacity to interrogate underlying assumptions due to structural inertia, implement these recommendations, allocating resources toward already-served populations. These allocations, in turn, justify and reinforce the political-economic logic that privileges formal, urban data collection over the documentation of informal and rural realities. The feedback loop between algorithmic bias, resource allocation, and governance is a classic co-evolutionary process (Kauffman, 1993).

This completes the self-perpetuating loop. Inequitable allocations reinforce the governance logic that produces incomplete data, which in turn trains biased models to recommend further inequitable allocations.

**Why Technical Solutions Are Insufficient**

This systemic analysis reveals why isolated technical refinements fail:

- Improved model architectures cannot correct for ontological exclusions arising from path-dependent data.
- Sophisticated explainability techniques cannot address legitimacy deficits when structural inertia decouples technical evaluation from governance metrics.
- Bias detection audits cannot reverse inequitable allocations already implemented under the guise of objective advice.
- Enhanced data collection that preserves existing schemas merely generates higher-resolution maps of the same partial reality.

Liang et al. (2022) identify the design and sculpting of data as the foremost challenge for trustworthy AI, as data critically determines model outcomes. Addressing this, however, demands more than technical refinement. It requires institutional transformation. We need to dismantle governance silos that produce fragmented data, build endogenous capacity to enable ownership over dependency, and embed justice criteria within optimization architectures from the outset.

This logic fundamentally shapes AI applications deployed within this system. They cannot transform it because they are designed to operate within it. They optimize and provide diagnostics for existing governance processes rather than enabling their structural reconstitution.

## 5.6 Conclusion: Requirements for Complexity-Aware, Ethically Grounded Architecture

We identified three interconnected challenges detailed in this chapter. Arthur's (1989) path dependence in data infrastructure, Hannan and Freeman's (1984) structural inertia in institutional capacity, and systematic ethical failures in algorithmic optimization constitute definitive requirements for any computational framework aspiring to support equitable and adaptive energy policy.

These are not temporary obstacles awaiting technical solutions. They are stable system states maintained by self-reinforcing dynamics rooted in institutional path dependencies (Casula, 2022), political economies of technological provision (Unwin, 2017), and the structural imperatives of bureaucratic legibility (Debnath and Mourshed, 2018; Nilashi et al., 2023). Path-dependent data infrastructure persists because it serves bureaucratic accountability; structural inertia perpetuates because it serves reliability requirements; algorithmic bias amplifies because models optimize their given objective functions with precision.

Therefore, we must design the BRIDGE-E framework from foundational principles to contend with these dynamics:

1. Disrupt Arthur's (1989) path dependence with epistemic inclusion.

The framework must possess the capacity to ingest, validate, and reason with heterogeneous knowledge forms (e.g., qualitative community narratives, participatory assessments, traditional knowledge documentation) that explicitly counter the erasures of official data schemas.

2. Dissolve Hannan and Freeman's (1984) structural inertia with capacity co-production.

Institutional embedding must precede technical deployment. This necessitates open-source architectures enabling local modification, hybrid teams pairing external expertise with permanent institutional staff, transparent documentation, and funding models that bind external capital to sustained national commitments. This approach operationalizes decolonial AI governance principles (Gelan et al., 2024) and navigates institutional path dependency (Casula, 2022) by designing systems that require, rather than merely permit, institutional co-evolution.

3. Counter systematic ethical failure with justice-by-design.

Equity metrics must function as constitutive optimization constraints, not post-hoc evaluations. This requires contestable objective functions that enable affected populations to challenge algorithmic priorities, transparent architectures adhering to regulatory mandates (Papadakis et al., 2024), and organizational structures that position AI as scaffolding for human deliberation. This operationalizes Dignum's (2019) responsibility-by-design and Jenkins et al.'s (2016) energy justice within the computational logic itself, forcing systems to reason explicitly about distributive outcomes.

These are not aspirational enhancements but existential requirements. Section 5.5 demonstrates that the failure to address any one dynamic sabotages the entire system through feedback mechanisms. Epistemic inclusion without capacity co-production generates sophisticated diagnostics that external actors consume, perpetuating dependency. Capacity co-production without justice-by-design creates an endogenous capability to optimize unjust objectives efficiently. Justice-by-design without epistemic inclusion (Fricker, 2007) merely refines optimization over partial ontologies, systematically excluding marginalized realities.

These requirements constitute a diagnostic framework. They predict that initiatives which fail to integrate epistemic inclusion, capacity co-production, and justice-by-design will be captured by the very dynamics they aim to overcome. Chapter 6 tests this prediction through empirical analysis. We examine various initiatives to validate whether the path dependence, structural inertia, and ethical failures theorized here manifest as expected in institutional practice. This validation converts the diagnostic framework into an empirically grounded blueprint. The resulting specifications, derived from observed patterns of failure, provide the definitive foundation for the BRIDGE-E framework developed in Part III.

## References


Arthur, W. B. (1989). Competing technologies, increasing returns, and lock-in by historical events. *The Economic Journal, 99*(394), 116–131. https://doi.org/10.2307/2234208

Casula, M. (2022). Implementing the transformative innovation policy in the European Union: How does transformative change occur in Member States? *European Planning Studies, 30*(11), 2178–2204. https://doi.org/10.1080/09654313.2021.2025345

de Bruijn, H., Warnier, M., & Janssen, M. (2022). The perils and pitfalls of explainable AI: Strategies for explaining algorithmic decision-making. *Government Information Quarterly, 39*(2), 101666. https://doi.org/10.1016/j.giq.2021.101666

Debnath, K. B., & Mourshed, M. (2018). Challenges and gaps for energy planning models in the developing-world context. *Nature Energy, 3*, 172–184. https://doi.org/10.1038/s41560-018-0095-2

Dignum, V. (2019). *Responsible artificial intelligence: How to develop and use AI in a responsible way*. Springer. https://doi.org/10.1007/978-3-030-30371-6

Floridi, L. (2013). *The ethics of information*. Oxford University Press. https://doi.org/10.1093/acprof:oso/9780199641321.001.0001

Fricker, M. (2007). *Epistemic injustice: Power and the ethics of knowing*. Oxford University Press. https://doi.org/10.1093/acprof:oso/9780198237907.001.0001

Gelan, A., et al. (2024). Decolonizing global AI governance: Assessment of the state of decolonized AI governance in Sub-Saharan Africa. *Royal Society Open Science, 11*(2), 231994. https://doi.org/10.1098/rsos.231994

Hannan, M. T., & Freeman, J. (1984). Structural inertia and organizational change. *American Sociological Review, 49*(2), 149–164. https://doi.org/10.2307/2095567

Holland, J. H. (1995). *Hidden order: How adaptation builds complexity*. Addison-Wesley.

Holland, J. H., *Emergence from chaos to order* (Oxford, 1998; online edn, Oxford academic, 31 Oct. 2023), https://doi.org/10.1093/oso/9780198504092.001.0001, accessed 20 Dec. 2025.

Jenkins, K., McCauley, D., Heffron, R., Stephan, H., & Rehner, R. (2016). Energy justice: A conceptual review. *Energy Research & Social Science, 11*, 174–182. https://doi.org/10.1016/j.erss.2015.10.004

Kauffman, S. A. (1993). *The origins of order: Self-organization and selection in evolution*. Oxford Academic. https://doi.org/10.1093/oso/9780195079517.001.0001

Kim, S., Wellstead, A. M., & Heikkila, T. (2023). Policy capacity and rise of data-based policy innovation labs. *Review of Policy Research, 40*, 341–362. https://doi.org/10.1111/ropr.12494

Klingbeil, A., Grützner, C., & Schreck, P. (2024). Trust and reliance on AI – An experimental study on the extent and costs of overreliance on AI. *Computers in Human Behavior, 160*, 108352. https://doi.org/10.1016/j.chb.2024.108352

Levinthal, D. A. (1997). Adaptation on rugged landscapes. *Management Science, 43*(7), 934–950. https://doi.org/10.1287/mnsc.43.7.934

Liang, W., Tadesse, G. A., Ho, D., et al. (2022). Advances, challenges and opportunities in creating data for trustworthy AI. *Nature Machine Intelligence, 4*, 669–677. https://doi.org/10.1038/s42256-022-00516-1

Lindgren, S., & Dignum, V. (2023). Chapter 14: Beyond AI solutionism: Toward a multi-disciplinary approach to artificial intelligence in society. In *Handbook of critical studies of artificial intelligence* (pp. 179–193). Edward Elgar Publishing. https://doi.org/10.4337/9781803928562.00019

Nilashi, M., Keng Boon, O., Tan, G., Lin, B., & Abumalloh, R. (2023). Critical data challenges in measuring the performance of sustainable development goals: Solutions and the role of big-data analytics. *Harvard Data Science Review, 5*(3). https://doi.org/10.1162/99608f92.545db2cf

Papadakis, T., Christou, I. T., Ipektsidis, C., Soldatos, J., & Amicone, A. (2024). Explainable and transparent artificial intelligence for public policymaking. *Data & Policy, 6*, e10. https://doi.org/10.1017/dap.2024.3

Sharma, M., Luthra, S., Joshi, S., & Kumar, A. (2022). Implementing challenges of artificial intelligence: Evidence from public manufacturing sector of an emerging economy. *Government Information Quarterly, 39*(4), 101624. https://doi.org/10.1016/j.giq.2021.101624

Starke, C., & Lünich, M. (2020). Artificial intelligence for political decision-making in the European Union: Effects on citizens' perceptions of input, throughput, and output legitimacy. *Data & Policy, 2*, e16. https://doi.org/10.1017/dap.2020.19

Unwin, T. (2017). *Reclaiming information and communication technologies for development*. Oxford University Press. https://doi.org/10.1093/oso/9780198795292.001.0001

Wellstead, A. M., Mechling, S. M., Carter, A., & Gofen, A. (2024). Artificial intelligence possibilities to improve analytical policy capacity: The case of environmental policy innovation labs and sustainable development goals. *Policy Design and Practice, 7*(4), 456–467. https://doi.org/10.1080/25741292.2024.2385118

# Chapter 6
# Empirical Validation: Testing the Complexity-Ethics Framework Through Institutional Practice

## 6.1 From Theory to Empirical Testing

As shown in Chap. 4 and 5, computational applications display complex adaptive systems (CAS) dynamics (Hannan & Freeman, 1984; Arthur, 1989; Kauffman, 1993; Holland, 1995, 1998) and perpetuate systematic ethical failures. Here, we test that diagnosis by asking: how do these dynamics manifest when well-resourced actors attempt computational governance innovation?

We examine a multilateral actor (UNDP Accelerator Labs, n.d.; UNDP, 2024), an international financial institution (World Bank ESMAP/DE4A), national energy efficiency programs, and a regional coordinator (ASEAN Centre for Energy). For each, we ask:

1. Which CAS dynamics (path dependence, structural inertia, co-evolutionary mismatch) prevent transformation?
2. Which ethical principles (epistemic justice, information ethics, and energy justice) are violated?
3. Do the implementation challenges theorized in Chap. 5 materialize?

The systematic analysis reveals a consistent pattern of failure. From this pattern, we derive the concrete specifications for the BRIDGE-E framework introduced in Chap. 7.


R. Bouderhem, *A Computational Framework for Energy Policy Coherence*, SpringerBriefs in Energy, https://doi.org/10.1007/978-3-032-22978-6_6

## 6.2 Case Studies: Four Institutional Contexts

### *6.2.1 UNDP Accelerator Labs and AI Hub: Multilateral Development Architecture*

**Context and Scope**

The United Nations Development Programme (UNDP) operates a portfolio of AI initiatives targeting sustainable development. Its Accelerator Labs network consists of 91 Lab teams across 115 countries, exploring machine learning (ML) applications for critical challenges, including energy efficiency. Recent deployments illustrate this technical focus. They include AI-based energy models developed with IBM to forecast global energy access (Jessen, 2024); the AI4SD open-source platform, launched with the UAE to provide AI models and datasets to policymakers across Arab States (UNDP Arab States, 2024); and a partnership in Rwanda with the AI-driven clean energy investment platform Nithio (UNDP, n.d.). The UNDP AI Hub also provides resources for building national AI capabilities (UNDP, 2024).

**Observed Manifestations of CAS Dynamics**

- Arthur's (1989) path dependence in data infrastructure. This dynamic is evident in the data architecture implicitly reinforced by capacity-building efforts. The initiatives confront a well-documented challenge. Populations most in need of policy intervention often lack systematic data representation (Hilbert, 2016). The strategic response, however, frequently prioritizes standardizing and interoperating existing formal data sources. This technical focus exemplifies path dependence. It invests in and reinforces the institutional data infrastructures whose ontological boundaries (e.g., categories for formal, grid-connected, revenue-generating activities) originally created the epistemic exclusion. Consequently, while data systems become more technically sophisticated, the fundamental gap remains unaddressed. There are no categorical structures to render informal, off-grid, or community-managed realities legible.
- Hannan and Freeman's (1984) structural inertia in institutions. This manifests as a persistent temporal (Levinthal & March, 1993) and co-evolutionary (Kauffman, 1993) mismatch. Externally designed initiatives follow constrained, output-oriented cycles (Unwin, 2017). However, the absorption of new capabilities follows the slower, path-dependent trajectories of national bureaucracies. This mismatch ensures intervention cycles conclude before genuine institutional transformation can take place. This leads to the dissipation of expertise and reinforced dependency documented in policy innovation labs (Kim et al., 2023).

**Observed Ethical Failures**

The logic of these initiatives systematically reproduces the ethical failures theorized in Chap. 5:

- Fricker's (2007) hermeneutical injustice. Populations in informal settlements, off-grid communities, and unprofitable service areas lack the categorical representation within official taxonomies required to make their energy needs and practices intelligible to policy systems.

- Floridi's (2013) informational entropy. By generating ever more voluminous, high-precision data about formally represented sectors, these systems actively degrade the overall information ecosystem's capacity to represent marginalized realities. They produce authoritative maps of a partial world.
- Divergence from responsibility-by-design. Equity only functions as an external objective. It is not as an architectural constraint determining admissible solutions. This contrasts with the principle of responsibility-by-design (Dignum, 2019), where equity and justice are constitutive parameters baked into the system's core logic from the outset.

**Validation of Chap. 5 Challenges**

The UNDP case provides empirical validation for the tripartite diagnosis:

1. Path-dependent data infrastructure. Data systems evolve toward greater formal interoperability and sophistication, yet the underlying ontological exclusion of nonconforming realities persists.
2. Structural inertia in capacity development. Capacity building creates "islands" of analytical skill but fails to generate the political capacity and architectural ownership required for endogenous transformation. This mirrors the "awareness-action gap" in AI governance, where, as Gelan et al. (2024) document, 80% of evaluated Sub-Saharan African countries remain "decolonization-aware" without substantive progress.
3. Systematic ethical failure in optimization. The computational logic remains decoupled from core justice imperatives. Equity is an aspirational goal appended to the system's outputs, not the constitutive logic governing its search for solutions.

### 6.2.2 *World Bank ESMAP and DE4A: International Financial Institution Architecture*

**Context and Scope**

The World Bank Group applies AI to energy governance through two primary, complementary channels. The Energy Sector Management Assistance Program (ESMAP) provides technical assistance and supports initiatives that leverage AI for energy policy analysis in developing countries (World Bank ESMAP, 2023). Concurrently, the Digital Economy for Africa (DE4A) initiative aims for universal digital enablement in Africa by 2030, aligning with the African Union's Digital Transformation Strategy (World Bank DE4A, 2023; African Union, 2020). This reflects a coordinated approach in line with principles of international policy coherence (World Bank IEG, 2023; OECD, n.d.).

**Observed Manifestations of CAS Dynamics**

- Arthur's (1989) path dependence in data standardization. The World Bank's initiatives address the profound "data quality, availability, and adequacy" challenges that Debnath and Mourshed (2018) identify as systematic deficiencies in energy planning for developing contexts. A central methodological response involves promoting technical data standardization to establish interoperable schemas, making national datasets machine-readable and comparable. This is essential for enabling analysis. However, this process inherently embeds and perpetuates specific ontological choices (decisions about what phenomena we categorize and measure) into the data infrastructure. This exemplifies path dependence. Investments in standardization reinforce the conceptual boundaries of existing formal data systems. As Nilashi et al. (2023) argue, the resulting data incompleteness for tracking Sustainable Development Goals is often structural, excluding entire dimensions of reality from representation. Thus, the pursuit of interoperability can advance technical coherence while simultaneously systematizing epistemic exclusions. This is particularly true regarding informal economies, off-grid systems, and community-managed resources that lack representation in the standardized schemas.
- Hannan and Freeman's (1984) structural inertia in operational procedures. The World Bank's operational model, refined over decades for fiduciary accountability, creates a powerful institutional trajectory that generates co-evolutionary mismatches (Kauffman, 1993) with client countries' governance systems. Project-based interventions successfully demonstrate technical feasibility and build individual expertise. However, as Sharma et al. (2022) document in their study of public sector AI implementation, such capacity often fails to institutionalize within permanent government structures. This gap aligns with Kim et al.'s (2023) finding that external initiatives can build analytical capacity (skills and tools) without developing the necessary political capacity—the institutional authority, legitimacy, and decision-rights required to transform policy processes endogenously.

**Observed Ethical Failures**

Such initiatives establish a fundamentally different ethical orientation from the principles of responsibility-by-design (Dignum, 2019) and the technical mandates for explainable AI in policymaking (Papadakis et al., 2024).

- The risk management approach positions equity as a mitigation constraint on technically pre-optimized solutions. In contrast, responsibility-by-design requires equity to be the generative objective, defining the space of computationally admissible solutions from the outset. The former permits solutions with high technical efficacy provided they manage negative social impacts; the latter would preclude the release of such solutions.
- This paradigm structurally decouples the optimization objectives of AI systems (e.g., cost-effectiveness, aggregate efficiency) from the core tenets of energy justice (Jenkins et al., 2016). Distributional equity, procedural fairness, and

recognition of diverse needs become external concerns to mitigate, not an internal logic governing system design.

**Validation of Chap. 5 Challenges**

The World Bank case provides empirical validation of the tripartite diagnosis:

1. Path-dependent data infrastructure. Investments drive data systems toward greater formal sophistication and interoperability, yet the epistemic boundaries of what constitutes valid data remain fixed, systematically excluding nonconforming realities.
2. Structural inertia in capacity development. The project-based delivery model generates demonstrable but temporary capabilities, perpetuating a cycle of dependency rather than fostering the sovereign, architectural ownership required for transformative governance.
3. Systematic ethical failure in optimization. The dominant ethical framework of risk mitigation ensures optimization occurs within a technocratic value space.

Consequently, justice is not a constitutive design principle but a post hoc corrective. It allows computational tools to reinforce existing socio-economic inequalities.

### *6.2.3 National Energy Efficiency Programs: Domestic Implementation Architecture*

**Context and Scope**

National energy efficiency programs represent cases where AI capabilities achieve higher degrees of operational integration. Machine learning (ML) applications, such as Non-Intrusive Load Monitoring (NILM) for disaggregating and analyzing household appliance consumption, demonstrate technical sophistication and direct policy relevance (Garcia et al., 2020). These systems operate on rich, historically accumulated datasets, primarily national utility consumption records, to generate actionable efficiency insights.

**Observed Manifestations of CAS Dynamics**

- Arthur's (1989) path dependence in utility data architecture. ML systems excel at analyzing the formal, grid-connected energy sector precisely because utility data architectures comprehensively capture that reality. Decades of utility operations, optimized for billing and grid management, have produced datasets that encode a specific ontology: metered consumption, load profiles, and customer billing information. This creates a path-dependent analytical ceiling. As Nepal et al. (2020) document in the context of the Hindu Kush Himalaya (HKH) region, "little is known about the regional energy consumption pattern and trends in the HKH due to data scarcity." This makes evidence-based policy difficult. The ML systems inherit this scarcity. Phenomena such as off-grid microgrids, community energy cooperatives, informal sector use, and suppressed demand in unserved areas lack the categorical structures through which utility databases represent energy. The data architecture determines the model's epistemic horizon.

- Hannan and Freeman's (1984) structural inertia manifests in the challenge of institutional integration. In practice, the advanced ML capabilities for national programs are frequently developed and delivered through external partnerships (e.g., with tech companies, academic institutions, or technical consultancies). The critical juncture is whether this externally sourced expertise catalyzes permanent institutional competency or remains a transient augmentation. This challenge directly relates to the core requirement for public sector AI transformation identified in the literature. It necessitates an "operational transformation" with "human resources and know-how as the spearhead" (Valle-Cruz & García-Contreras, 2025). The failure to convert external technical inputs into endogenous institutional capacity leaves innovation vulnerable to dissipation. Consequently, it reproduces the structural inertia of governance systems.

**Observed Ethical Failures**

- Distributional injustice in optimization. ML systems designed to maximize aggregate energy reduction will systematically recommend interventions with the largest absolute savings (e.g., industrial facilities, commercial buildings, and high-consumption residential areas). As Mehmood et al. (2019) review, AI applications in building energy efficiency focus on optimizing these technically legible domains. Without explicit constraints prioritizing distributive justice (Jenkins et al., 2016), optimization inherently favors populations already well-represented in the training data.
- Epistemic (testimonial) injustice (Fricker, 2007). The bias is not only in the algorithm but also in the foundational political choices that define the system. The objective function (aggregate reduction over equity-weighted access), the curation of training data (historical patterns reflecting past infrastructural exclusions), and the success metrics (technical kilowatt-hour savings over equitable access gains) perpetuate epistemic injustice (Fricker, 2007). These applications systematically discount the energy needs and realities of marginalized populations. Their experiences lack the requisite credibility within the data structures that inform algorithmic recommendations.

**Validation of Chap. 5 Challenges**

This validates the persistent triad of challenges:

1. Path-dependent data infrastructure. The utility data paradigm constrains what AI models can see and optimize for, creating a structurally incomplete knowledge base for policy.
2. Structural inertia in capacity development. The dependency on external partnerships for advanced ML creates fragility. Unless specifically countered, this leads to capability dissipation rather than endogenous institutional transformation.
3. Systematic ethical failure in optimization. Operating on ontologically biased and justice-blind data, these initiatives will produce inequitable policy

recommendations. This will reinforce existing inequalities under a veneer of technical neutrality.

### 6.2.4 *ASEAN Centre for Energy: A Theoretical Probe into Regional Coordination*

**Context and Analytical Focus**

The ASEAN Centre for Energy (ACE) operates at the apex of coordination complexity. Its task is to align policy across 11 sovereign states with "different energy needs, resources, and developmental priorities" (Leong, 2023). Moving from empirical case studies to theoretical analysis, this point does not examine a specific AI deployment. Instead, it uses ACE's well-documented institutional position and coordination challenges as a foundational scenario to probe a critical question. If ASEAN implements AI governance at a regional scale, how would the systemic dynamics identified in our framework manifest and intensify? This theoretical extrapolation, grounded in a real-world coordination mandate, reveals the compounded challenges any regional computational framework would face.

**Theoretical Manifestations of CAS Dynamics**

- Compounded path dependence (Arthur, 1989). Regional AI governance would necessitate integrating disparate national datasets. The logical technical solution (data harmonization) would seek interoperability among existing national data infrastructures. This process would operationalize compounded path dependence. ASEAN will achieve technical coherence by reinforcing and interlocking the pre-existing, path-dependent data ontologies of each member state. The incentive structure would inherently favor improving formal interoperability over addressing each nation's underlying epistemic exclusions, systematically locking in a region-wide ontological bias.
- Intensified structural inertia (a logical extension). The structural inertia theorized by Hannan and Freeman (1984) within a single bureaucracy would be multiplied at the regional scale. Coordination would depend on variable donor funding or member states' financial contributions and synchronized political commitment across multiple sovereign entities with divergent fiscal cycles and priorities. Capabilities would likely advance through fragile, time-bound projects rather than as permanent regional capacity. This creates a near-insurmountable co-evolutionary mismatch (Kauffman, 1993), demanding the simultaneous, aligned transformation of multiple national bureaucracies. This is a hallmark of structural inertia.

**Theoretical Ethical Failures and Legitimacy Challenges**

- Spatial injustice from heterogeneous preconditions. The implementation of regional AI would occur across vastly different national contexts. Research in other regions, such as the MENA area, shows a causal link between energy

efficiency and preconditions like AI readiness and GDP per capita (Hossin et al., 2023). This finding logically predicts structurally reinforced spatial inequality in ASEAN. Models trained on data from more advanced members would generate recommendations ill-suited or disadvantageous to less developed members, enacting algorithmic distributional injustice at a regional scale.
- The inevitable confrontation with value pluralism. ASEAN members have legitimately different policy priorities (e.g., energy security vs. climate mitigation). A regional AI system optimizing for an aggregate metric would inevitably disadvantage states prioritizing alternative values. This directly engages the legitimacy crisis identified by Starke and Lünich (2020). Their research found that citizens perceive purely algorithmic systems as illegitimate sole decision-makers. In the ASEAN context, where members have legitimate value pluralism, an AI system optimizing for an aggregate regional metric would inherently override local political preferences, triggering and amplifying this illegitimacy crisis. Furthermore, as de Bruijn et al. (2022) establish, technical explainability is insufficient in such "wicked" problems where the core conflict is over normative criteria and legitimate objectives, not computational mechanics.

**Validation of the Framework's Predictive Logic**
This validates our framework's internal logic and predictive power by demonstrating that its core dynamics would not dissipate but intensify at the regional level:

1. Path dependence compounds: national exclusions would aggregate into a region-wide ontological blind spot.
2. Structural inertia multiplies: coordination costs and co-evolutionary mismatches would magnify institutional resistance to transformation.
3. Ethical challenges become existential: distributional injustice would manifest as spatial inequality, and conflicts over values would create fundamental legitimacy deficits that technical solutions cannot resolve.

This analysis confirms that the challenges identified through our framework are not incidental but structural properties of scaling computational governance across institutional and sovereign boundaries.

## 6.3 Architectural Requirements Derived from Empirical Evidence

The consistent pattern of failure across these diverse initiatives validates the explanatory power of the complexity-ethics framework. Despite varying resources and intentions, there is no transformative change due to Arthur's (1989) path dependence, Hannan and Freeman's (1984) structural inertia, and persistent co-evolutionary mismatches (Kauffman, 1993). Concurrently, Fricker's (2007) epistemic injustice, Floridi's (2013) informational entropy, and violations of Jenkins et al.'s (2016) energy justice principles persistently materialize. This systematic recurrence demonstrates these are not implementation accidents but structural inevitabilities of the current computational governance paradigm.

The initiatives analyzed function as technical augmentations. They are not the catalytic transformers required to dismantle the governance-data-inequality nexus. This diagnostic conclusion resonates with a growing consensus on the need for a fundamental architectural shift. For instance, Danish and Senjyu (2023) argue that realizing AI's potential demands "an AI-driven policy framework that aligns with the circular economy business model." Our analysis provides the specification for such a framework. It must directly counter the systemic dynamics and ethical failures empirically documented here.

From this evidence, we derive four foundational requirements for any computational framework aspiring to enable coherent and just energy governance.

**Requirement 1: Operationalizing Fricker's Epistemic Justice**

- Observed failure: initiatives prioritize standardizing and interoperating existing formal data. This reinforces the path-dependent choices optimized for bureaucratic legibility that exclude tacit, community-held, and informal knowledge (Nilashi et al., 2023; Nepal et al., 2020). This actively produces hermeneutical injustice (Fricker, 2007) and informational entropy (Floridi, 2013).
- Architectural requirement: frameworks must actively counter Arthur's (1989) path dependence in data ontologies through mechanisms designed for epistemic inclusion. This necessitates systems capable of ingesting and validating heterogeneous knowledge forms like qualitative narratives, participatory mapping data, and locally defined indicators. Here, we could use knowledge graph structures that represent competing ontological claims rather than enforcing singular taxonomies. Validation must assess epistemic adequacy (Does this represent all affected populations?) alongside technical validity. Epistemic justice and information ethics must be constitutive design constraints.

**Requirement 2: Addressing Hannan and Freeman's Structural Inertia**

- Observed failure: current initiatives focus on delivering tools or building individual human capital (analytical and operational capacity) without redesigning the institutional routines, incentives, and decision-rights that govern their use. As Kim et al. (2023) demonstrate in their study of data-based policy innovation labs, such external initiatives often succeed in building analytical and operational capacity but "do little to enhance political capacity." They do not have the institutional authority and legitimacy required for endogenous transformation. Consequently, capacity building remains additive and externally dependent rather than transformative. This political capacity gap is exacerbated by the implementation challenges documented within public sector institutions themselves. Sharma et al. (2022), studying India's public manufacturing sector, identify these as "poor data quality, managers' lack of understanding of cognitive technologies... and expensive technologies." These are not mere technical hurdles but symptoms of the endogenous institutional barriers that represent an organization's structural defenses against disruptive change. This structural inertia, combined with temporal mismatches between short-term intervention cycles

and long-term institutional learning, also prevents genuine integration and violates decolonial governance principles (Gelan et al., 2024).
- Architectural requirement: frameworks must overcome structural inertia through institutional co-evolution. This requires moving beyond technical delivery to enable what Valle-Cruz and García-Contreras (2025) identify as an "operational transformation" with "human resources and know-how as the spearhead." Endogenous capacity co-production must structure the deployment process. It requires explicit roadmaps for transferring architectural ownership. It necessitates creating hybrid roles to bridge technical expertise and governance practice. It demands aligning computational system evolution with national budgetary and political cycles, not external donor timelines. Finally, it requires designing for path creation rather than accepting path dependence.

**Requirement 3: Operationalizing Jenkins et al.'s Energy Justice and Dignum's Responsible AI**

- Observed failure: risk management treats equity as a post-design concern rather than an objective for optimization from inception. This reflects the limitations of current impact assessment paradigms. Mantelero (2018) analyzes frameworks like the GDPR's Data Protection Impact Assessment for having "a narrow focus—centred on data quality and data security" that fails to adequately address broader societal impacts. He argues these models are "too closely focused on data processing" and "do not adequately highlight ethical and social issues" (Mantelero, 2018). They effectively relegate concerns like algorithmic discrimination and distributional injustice to peripheral considerations. Consequently, optimization objectives center on aggregate efficiency, cost, or technical performance, biasing outcomes toward populations already over-represented in training data. This violates the distributional, procedural, and recognition dimensions of Jenkins et al.'s (2016) energy justice and perpetuates Fricker's (2007) testimonial injustice.
- Architectural requirement: frameworks must transcend reactive risk management through energy justice and responsibility-by-design principles (Dignum, 2019). This requires encoding equity as a primary, generative constraint. This goes beyond the assessment-focused approach of models like the Human Rights, Ethical and Social Impact Assessment (HRESIA) proposed by Mantelero (2018). While such models crucially expand the evaluative framework beyond data security, they remain diagnostic. Our requirement is architectural. Equity metrics must be constitutive. Distributive equity (e.g., rural-urban parity), procedural justice (affected population contestability), and recognition (including marginalized ontologies) must function as hard constraints within objective functions. The system must then identify governance approaches that satisfy these equity constraints, optimizing only within justice-defined boundaries. Equity becomes the guiding principle determining admissible solutions made by computational models.

**Requirement 4: Respecting Complexity Dynamics in Governance Design**

- Observed failure: initiatives deploy AI as advisory tools within unchanged governance processes. They implicitly treat policymaking as an optimization problem with computable solutions. This fails to respect policymaking as a deliberative process within complex adaptive systems (CAS) (Holland, 1995), which involve multiple legitimate values, irreducible uncertainties, and co-evolutionary dynamics.
- Architectural requirement: we must facilitate interaction between human stakeholders, domain knowledge, and computational models within CAS dynamics, not seek to automate decisions. This requires structured protocols for deliberative sense-making about which objectives merit optimization; simulation environments to stress-test policies against diverse value priorities, revealing unintended consequences; and feedback mechanisms where real-world policy outcomes refine model parameters. The role of computation is to support more intelligent governance conversations, respecting Kauffman's (1993) and Levinthal's (1996) complexity dynamics. This aligns with the finding of Starke and Lünich (2020) that citizens perceive algorithmic systems as illegitimate sole decision-makers. The framework must therefore preserve and enhance, not replace, human political judgment and legitimacy.

## 6.4 Empirically Validated Specifications

These institutional analyses confirm the complexity-ethics analytical lens. Arthur's (1989) path dependence, Hannan and Freeman's (1984) structural inertia, and co-evolutionary mismatches (Kauffman, 1993; Holland, 1995) prevent transformation even with substantial resources and explicit equity commitments. Fricker's (2007) epistemic injustice, Floridi's (2013) informational entropy, and Jenkins et al.'s (2016) energy justice violations persistently materialize. This systematic recurrence demonstrates that the failures are structural features of the current applications' design and deployment logic.

The requirements define a problem-space current applications cannot solve through incremental improvements. Sophisticated algorithms, additional training, or finer-grained bias audits fail because they operate within existing boundaries. They are unable to overcome path dependence, break through structural inertia, or resolve co-evolutionary mismatches.

An effective response must emerge from first principles to satisfy this empirically validated specification. Addressing Arthur's path dependence through mechanisms operationalizing Fricker's (2007) epistemic justice and Floridi's (2013) information ethics; overcoming Hannan and Freeman's (1984) structural inertia through institutional transformation building sovereignty; implementing Jenkins et al.'s (2016) energy justice and Dignum's (2019) responsibility-by-design by encoding equity as a primary constraint; and respecting complexity dynamics through deliberative protocols maintaining democratic legitimacy.

With these requirements established, we turn to their instantiation. Chapter 7 introduces the BRIDGE-E framework, a computational architecture where each component is a direct response to the path dependence, structural inertia, and ethical failures diagnosed in Chaps. 4, 5 and 6.

## References


African Union. (2020). *Digital transformation strategy for Africa (2020–2030)*. https://au.int/en/documents/20200518/digital-transformation-strategy-africa-2020-2030

Arthur, W. B. (1989). Competing technologies, increasing returns, and lock-in by historical events. *The Economic Journal, 99*(394), 116–131. https://doi.org/10.2307/2234208

Danish, M. S. S., & Senjyu, T. (2023). Shaping the future of sustainable energy through AI-enabled circular economy policies. *Circular Economy, 2*(2), 100040. https://doi.org/10.1016/j.cec.2023.100040

de Bruijn, H., Warnier, M., & Janssen, M. (2022). The perils and pitfalls of explainable AI: Strategies for explaining algorithmic decision-making. *Government Information Quarterly, 39*(2), 101666. https://doi.org/10.1016/j.giq.2021.101666

Debnath, K. B., & Mourshed, M. (2018). Challenges and gaps for energy planning models in the developing-world context. *Nature Energy, 3*, 172–184. https://doi.org/10.1038/s41560-018-0095-2

Dignum, V. (2019). *Responsible artificial intelligence: How to develop and use AI in a responsible way*. Springer. https://doi.org/10.1007/978-3-030-30371-6

Floridi, L. (2013). *The ethics of information*. Oxford University Press. https://doi.org/10.1093/acprof:oso/9780199641321.001.0001

Fricker, M. (2007). *Epistemic injustice: Power and the ethics of knowing*. Oxford University Press. https://doi.org/10.1093/acprof:oso/9780198237907.001.0001

Garcia, F. D., Souza, W. A., Diniz, I. S., et al. (2020). NILM-based approach for energy efficiency assessment of household appliances. *Energy Informatics, 3*, 10. https://doi.org/10.1186/s42162-020-00131-7

Gelan, A., et al. (2024). Decolonizing global AI governance: Assessment of the state of decolonized AI governance in Sub-Saharan Africa. *Royal Society Open Science, 11*(2), 231994. https://doi.org/10.1098/rsos.231994

Hannan, M. T., & Freeman, J. (1984). Structural inertia and organizational change. *American Sociological Review, 49*(2), 149–164. https://doi.org/10.2307/2095567

Hilbert, M. (2016). Big data for development: A review of promises and challenges. *Development Policy Review, 34*, 135–174. https://doi.org/10.1111/dpr.12142

Holland, J. H. (1995). *Hidden order: How adaptation builds complexity*. Addison-Wesley.

Holland, J. H. (1998). *Emergence from chaos to order*. Oxford University Press. https://doi.org/10.1093/oso/9780198504092.001.0001. Accessed 20 Dec. 2025.

Hossin, M. A., Alemzero, D., Wang, R., Kamruzzaman, M. M., & Mhlanga, M. N. (2023). Examining artificial intelligence and energy efficiency in the MENA region: The dual approach of DEA and SFA. *Energy Reports, 9*, 4984–4994. https://doi.org/10.1016/j.egyr.2023.03.113

Jenkins, K., McCauley, D., Heffron, R., Stephan, H., & Rehner, R. (2016). Energy justice: A conceptual review. *Energy Research & Social Science, 11*, 174–182. https://doi.org/10.1016/j.erss.2015.10.004

Jessen, J. (2024, September 19). *IBM & UNDP: New energy models for global energy access*. Energy Digital. https://energydigital.com/articles/ibm-and-undp-launch-free-interactive-energy-models

Kauffman, S. A. (1993). *The origins of order: Self-organization and selection in evolution*. Oxford Academic. https://doi.org/10.1093/oso/9780195079517.001.0001

Kim, S., Wellstead, A. M., & Heikkila, T. (2023). Policy capacity and rise of data-based policy innovation labs. *Review of Policy Research, 40*, 341–362. https://doi.org/10.1111/ropr.12494

Leong, W. Y. (2023). Investigating and Enhancing Energy Management Policy and Strategies in ASEAN. In *2023 Asia meeting on environment and electrical engineering (EEE-AM), Hanoi, Vietnam*, pp. 1–6. https://doi.org/10.1109/EEE-AM58328.2023.10395060

Levinthal, D. (1996). Learning and Schumpeterian dynamics. In G. Dosi & F. Malerba (Eds.), *Organization and strategy in the evolution of the enterprise*. Palgrave Macmillan. https://doi.org/10.1007/978-1-349-13389-5_2

Levinthal, D. A., & March, J. G. (1993). The myopia of learning. *Strategic Management Journal, 14*, 95–112. https://doi.org/10.1002/smj.4250141009

Mantelero, A. (2018). AI and big data: A blueprint for a human rights, social and ethical impact assessment. *Computer Law & Security Review, 34*(4), 754–772. https://doi.org/10.1016/j.clsr.2018.05.017

Mehmood, M. U., Chun, D., Zeeshan, Han, H., Jeon, G., & Chen, K. (2019). A review of the applications of artificial intelligence and big data to buildings for energy-efficiency and a comfortable indoor living environment. *Energy and Buildings, 202*, 109383. https://doi.org/10.1016/j.enbuild.2019.109383

Nepal, R., Sharma, B., & al Irsyad, M. I. (2020). Scarce data and energy research: Estimating regional energy consumption in complex economies. *Economic Analysis and Policy, 65*, 139–152. https://doi.org/10.1016/j.eap.2019.12.002

Nilashi, M., Keng Boon, O., Tan, G., Lin, B., & Abumalloh, R. (2023). Critical data challenges in measuring the performance of sustainable development goals: Solutions and the role of big-data analytics. *Harvard Data Science Review, 5*(3). https://doi.org/10.1162/99608f92.545db2cf

OECD. (n.d.). *Development co-operation evaluation and effectiveness, evaluation criteria*. https://www.oecd.org/en/topics/sub-issues/development-co-operation-evaluation-and-effectiveness/evaluation-criteria.html

Papadakis, T., Christou, I. T., Ipektsidis, C., Soldatos, J., & Amicone, A. (2024). Explainable and transparent artificial intelligence for public policymaking. *Data & Policy, 6*, e10. https://doi.org/10.1017/dap.2024.3

Sharma, M., Luthra, S., Joshi, S., & Kumar, A. (2022). Implementing challenges of artificial intelligence: Evidence from public manufacturing sector of an emerging economy. *Government Information Quarterly, 39*(4), 101624. https://doi.org/10.1016/j.giq.2021.101624

Starke, C., & Lünich, M. (2020). Artificial intelligence for political decision-making in the European Union: Effects on citizens' perceptions of input, throughput, and output legitimacy. *Data & Policy, 2*, e16. https://doi.org/10.1017/dap.2020.19

UNDP. (2024). *G7 Italia, AI hub for sustainable development*. https://www.undp.org/sites/g/files/zskgke326/files/2024-07/ai_hub_report_digital.pdf

UNDP. (n.d.). *Sustainable energy hub, climate aggregation platform*. https://www.undp.org/climate-aggregation-platform/ai-enabled-financing-scale-energy-access

UNDP Accelerator Labs. (n.d.). https://www.undp.org/acceleratorlabs

UNDP Arab States. (2024, September 24). *UNDP and UAE launch a new platform leveraging the power of artificial intelligence for accelerating progress toward the Sustainable Development Goals in the Arab region*. https://www.undp.org/arab-states/press-releases/undp-and-uae-launch-new-platform-leveraging-power-artificial-intelligence-accelerating-progress-towards-sustainable

Unwin, T. (2017). *Reclaiming information and communication Technologies for Development*. Oxford University Press. https://doi.org/10.1093/oso/9780198795292.001.0001

Valle-Cruz, D., & García-Contreras, R. (2025). Towards AI-driven transformation and smart data management: Emerging technological change in the public sector value chain. *Public Policy and Administration, 40*(2), 254–275. https://doi.org/10.1177/09520767231188401

World Bank DE4A. (2023). *DE4A Newsletter 2023*. https://thedocs.worldbank.org/en/doc/ad89cf4cba9d3daeca993991923160bd-0400012023/original/DE4A-Newsletter-Fall-2023.pdf

World Bank ESMAP. (2023). *Energy sector management assistance program, annual report 2023*. https://www.esmap.org/sites/default/files/esmap-files/World%20Bank%20ESMAP%20AR%202023_DIGITAL.pdf

World Bank IEG. (2023). *World Bank Group support to demand-side energy efficiency*. Independent Evaluation Group/World Bank. https://documents1.worldbank.org/curated/en/099432403022338116/pdf/SECBOS0352c8d90dd0bf8a079a63fc9212b.pdf

# Part III
# The BRIDGE-E Framework

The Bias-Resistant Integration of Data, Governance, and Equity for Energy (BRIDGE-E) framework operationalizes the theoretical synthesis of complexity science and ethical frameworks developed in Part I. Its purpose is to satisfy the four requirements derived from the empirical analysis in Part II. We position BRIDGE-E as a testable theoretical proposition.

In Chap. 7, we detail the integrated components of BRIDGE-E. The data harmonization module counters path dependence by employing knowledge graphs that represent competing ontological claims and community-validated synthetic data generation to expand hermeneutical resources. The coherence assessment engine addresses systemic evaluation failures by encoding equity as a primary constraint, utilizing transformer-based semantic analysis, graph neural networks for cross-sectoral interaction modeling, and explainable AI for transparency. The context-sensitive recommendation system overcomes structural inertia through human-in-the-loop protocols that mandate stakeholder participation from inception. This chapter provides implementation specifications for resource-constrained environments and establishes a rigorous validation protocol to test whether the framework successfully operationalizes the architectural requirements.

In Chap. 8, we synthesize our argument to advance a new paradigm for governance in algorithmic societies. We demonstrate that traditional governance is structurally inadequate for systems characterized by emergence, path dependence, and co-evolution. The chapter articulates the necessary evolution—epistemic, institutional, and ethical—toward governance that is itself a complex adaptive system. BRIDGE-E serves as a theoretical exemplar of this evolution. It illustrates how computational systems can support rather than subvert democratic deliberation. The conclusion outlines a research agenda for governing complexity through iterative experimentation and institutional learning.

Part III thus completes the progression from identifying the crisis of governance to proposing both a specific tool and a fundamental reconceptualization of governance itself.

learning processes. It requires explicit roadmaps for transferring architectural ownership and designing for path creation rather than accepting path dependence.

**Requirement 3 (energy justice as primary constraint):** frameworks must operationalize Jenkins et al.'s (2016) energy justice and Dignum's (2019) responsibility-by-design by encoding equity as a primary constraint. Distributive equity, procedural justice, and recognition must function as hard constraints in objective functions, with equity as generative grammar determining computationally admissible solutions.

**Requirement 4 (complexity-aware governance):** frameworks must design interaction between human stakeholders, domain knowledge, and computational models within CAS dynamics rather than automating decisions. This requires structured protocols for deliberative sense-making, simulation environments stress-testing policies against diverse value priorities, and feedback mechanisms. The goal is to create genuine co-evolution between computational systems and governance processes.

### *7.1.2 BRIDGE-E as Operationalization of Established Theory*

BRIDGE-E translates the complexity-ethics lens and the diagnostic power of the governance-data-inequality nexus into a concrete architecture for what political science now examines as "AI as Governance" (Farrell, 2025). This framework engages directly with the field of algorithmic governance, understood as the use of digital technologies to produce specific forms of "social ordering" (Katzenbach & Ulbricht, 2019). As shown, "algorithmic governance is multiple, contingent and contested. It takes different forms in different contexts and jurisdictions, and it is shaped by interests, power, and resistance" (Katzenbach & Ulbricht, 2019).

In line with the call to move beyond AI solutionism through multidisciplinary integration (Lindgren & Dignum, 2023), BRIDGE-E steers algorithmic governance away from opaque, inertial, or unjust outcomes through the principles of complexity-aware ethics. Each component addresses the CAS dynamics and ethical failures documented in Chaps. 4, 5 and 6.

The framework applies Holland's (1995, 1998) principles of complex adaptive systems, Arthur's (1999) complexity economics, and Kauffman's (1993) co-evolutionary dynamics to policy environments. It engages directly with the critical discourse on algorithmic regulation (Yeung, 2018) by designing a system that embeds democratic deliberation and ethical constraints at its core. BRIDGE-E aims to harness computational power without ceding legitimate governance authority.

This builds upon the insight of Hayek (1945) that the knowledge required for societal coordination is inherently dispersed, local, and often tacit. This makes purely centralized or technocratic optimization impossible. As Sterman (2000) establishes for system dynamics, effective governance instruments must not only acknowledge path dependence but also identify evolutionary pathways, respect distributed knowledge, and treat coherence as an emergent property of well-orchestrated systemic interactions.

BRIDGE-E implements four ethical frameworks through specific technical mechanisms:

- Fricker's (2007) **epistemic justice:** we address (Sect. 7.2.1) testimonial injustice through structured validation channels and hermeneutical injustice through synthetic data generation validated by marginalized populations. This creates conceptual vocabulary for marginalized realities previously absent from official data schemas.
- Floridi's (2013) **information ethics:** we treat policy incoherence as informational entropy degrading the governance infosphere. Data harmonization functions as an entropy reduction mechanism creating semantic interoperability. We also implement Floridi and Taddeo's (2016) data ethics by designing computational spaces where diverse stakeholder perspectives can be reconciled.
- Dignum's (2019) **responsible AI:** explainable AI (XAI) components and continuous monitoring protocols concretize transparency, accountability, and auditability. BRIDGE-E's accountability mechanisms include contestability-by-design (Ananny and Crawford, 2018), structured answerability (Novelli et al., 2024), and maintained audit trails. These mechanisms are rooted in the key dimensions of algorithmic accountability (Wieringa, 2020), creating systemic rather than token accountability.
- Jenkins et al.'s (2016) **energy justice:** the context-sensitive recommendation module (Sect. 7.2.3) embeds distributional equity, procedural justice, and recognition as constitutive optimization constraints rather than post-hoc evaluation criteria.

This integration aims to address a critical gap. This is what Mittelstadt (2019) identifies as the lack of proven methods for translating ethical principles into practice. BRIDGE-E implements the core ethical principles Jobin et al. (2019) identify through specific technical mechanisms.

## 7.2 BRIDGE-E Architectural Components

The BRIDGE-E framework comprises three integrated computational modules, each designed to address specific empirically validated requirements from Chap. 6.

### *7.2.1 Data Harmonization Module: Addressing Path Dependence*

**Theoretical Foundation**

Chapter 5 documented how current computational tools rely on datasets encoding partial ontologies, systematically excluding off-grid systems, community cooperatives, and informal sector use for instance. As Debnath and Mourshed (2018) establish, data scarcity in developing contexts is often ontological. Entire dimensions lack categorical representation rather than missing values in existing categories.

**Architectural Response**

The data harmonization module pursues Fricker's (2007) hermeneutical justice through mechanisms enabling epistemic inclusion.

1. **Heterogeneous data integration:** the module employs complementary techniques to extract and structure data from diverse sources. For numerical energy consumption analytics, big data methods from smart grid applications provide a foundation (Zhang et al., 2018). For textual and regulatory data (government policies, international agreements, local inputs), natural language processing and knowledge graph techniques enable the extraction and organization of entities and relationships. These techniques structured complex domain-specific information such as hazardous chemical regulations and accident records (Zheng et al., 2021). To mitigate dataset bias and ensure transparency, the provenance and characteristics of all textual corpora should use frameworks such as Data Statements for NLP (Bender and Friedman, 2018). This integrated knowledge base prepares the data for subsequent analysis of policy trends and misalignments in the coherence assessment engine.
2. **Knowledge graph architecture:** moving beyond databases optimized for query efficiency, the module creates knowledge ecosystems designed for epistemic justice. Knowledge graphs represent competing ontological claims rather than enforcing singular taxonomies. This enables systems to reason with qualitative narratives from community consultations, participatory mapping data, and locally defined indicators alongside formal policy documents.
3. **FAIR principles integration:** the framework implements the FAIR Guiding Principles (Wilkinson et al., 2016) for policy document management, ensuring findability, accessibility, interoperability, and reusability. This addresses fragmentation challenges documented across developing-country implementations (Raza et al., 2024).
4. **Synthetic data generation with community validation:** where gaps exist in formal data, the module employs synthetic data generation to address representation bias. However, as Deslandes Whitney and Norman (2024) establish, synthetic data risks "diversity-washing." This creates a false appearance of representation and may circumvent data subject consent. BRIDGE-E mitigates these risks through community validation protocols where affected populations participate actively in synthetic data generation. As demonstrated, marginalized populations must shape algorithmic categories from inception rather than ticking a box in pre-existing schemas.

**Empirical Validation**

Deep learning techniques demonstrate feasibility. Moore et al. (2022) used transfer learning to predict molecular properties for organic photovoltaics applications, validating synthetic data augmentation approaches. The module's capacity to supplement incomplete datasets while maintaining contextual accuracy addresses the data scarcity documented by Nepal et al. (2020) and Nilashi et al. (2023).

**How This Addresses Path Dependence**

The module addresses Arthur's (1989) path dependence in data ontologies with its capacity to ingest heterogeneous knowledge forms and validate them through community participation. Rather than reinforcing schemas optimized for bureaucratic legibility, it creates new pathways where marginalized realities gain categorical representation. Governance infospheres (Floridi, 2013) should promote flourishing through inclusive representation rather than generate entropy through systematic exclusion.

### *7.2.2 Coherence Assessment Engine: Operationalizing Complexity Dynamics and Justice Constraints*

**Theoretical Foundation**

Chapter 4 documented how current applications optimize technical metrics (prediction accuracy, computational efficiency) while remaining structurally decoupled from governance metrics (distributional equity, procedural justice, democratic accountability). This is the concrete manifestation of Hannan and Freeman's (1984) structural inertia in evaluation systems.

**Architectural Response**

BRIDGE-E embeds governance imperatives through three integrated mechanisms:

1. **Semantic policy alignment analysis:** we design transformer-based NLP models to assess textual consistency between national and regional policies across sectors such as energy, environment, and social policy. These sectors represent a well-documented governance challenge (Aagaard, 2015). Zhang and El-Gohary (2023) show transformer models can successfully align technical specifications with regulatory texts for compliance checking. Similarly, Zheng et al. (2022) achieve 90.1% accuracy in rule interpretation using knowledge-driven NLP frameworks. Beyond technical regulations, Firebanks-Quevedo et al. (2022) validate that NLP can analyze complex policy documents (forestry incentives) with 90–94% accuracy, reducing analysis from weeks to minutes and revealing systemic patterns.
2. **Cross-sectoral impact modeling:** Graph Neural Networks (GNNs) analyze interactions between policies across sectors (energy, transportation, environment), mapping relationships to detect conflicting policy directives. Recent AI-for-policy research demonstrates GNNs can successfully extract policy-relevant insights from complex networks, such as identifying key innovation ecosystem nodes to inform policy design (Zhu, 2025). By treating policy systems as networks where interactions generate emergent properties, this operationalizes

complexity science principles (Holland, 1995, 1998; Kauffman, 1993), revealing conflicts and synergies invisible to siloed analysis.

3. **Fairness and bias detection with explainability:** the module integrates explainable AI (XAI) techniques to assess whether policies disproportionately favor certain regions or sectors. Mangla et al. (2024) demonstrate this capability using SHapley Additive exPlanations (SHAP) to quantify how climatic factors influence greenhouse gas emissions, generating actionable policy insights. This validates XAI as a functional component for ensuring algorithmic reasoning transparency in policy assessment.

**Embedding Energy Justice as a Primary Constraint**

Rather than treating equity as an external consideration during technical optimization, equity constraints structurally determine the solution space. Distributive equity (rural-urban parity, income-group outcomes), procedural justice (affected population contestability), and recognition (marginalized ontologies) function as constitutive constraints in objective functions. This renders inequitable solutions computationally excluded from the optimization search space. This approach directly embeds Jenkins et al.'s (2016) energy justice within BRIDGE-E.

**How This Addresses Structural Inertia and Complexity Dynamics**

The framework encodes governance metrics directly into computational reward structures, rather than treating them as external evaluation criteria. By doing so, the engine counteracts the manifestation of Hannan and Freeman's (1984) structural inertia within algorithmic evaluation systems. The GNN-based cross-sectoral modeling embodies Kauffman's (1993) and Levinthal's (1997) insights about co-evolution on dynamic fitness landscapes. This enables the identification of emergent conflicts from policy interactions across evolving configurations (Holland, 1995, 1998).

### 7.2.3 *Context-Sensitive Recommendation System: Addressing Structural Inertia Through Deliberative Co-Validation*

**Theoretical Foundation**

As documented in Chap. 6, various initiatives consistently deploy AI as advisory tools within unchanged governance processes. Consequently, they will fail to build endogenous institutional capacity. They treat policymaking as an optimization problem, rather than a deliberative process within complex adaptive systems.

**Architectural Response**

The context-sensitive recommendation system implements complexity-aware, ethical governance through human-in-the-loop (HITL) protocols:

1. Structured deliberative co-validation: drawing from theories of participatory governance (Fischer, 2012) and deliberative democracy (Gutmann and Thompson, 2004), the HITL protocol creates structured spaces for contestation

and refinement of AI-generated recommendations. Debnath et al. (2025) demonstrate HITL's necessity and effectiveness in climate action. AI combined with HITL frameworks leverages digital data. This combination also integrates human knowledge for ethical, context-aware decisions, balancing efficiency with diverse cultural values and reducing biases. This validates HITL as essential for legitimate, implementable policy recommendations (Debnath et al., 2025), aligning with broader principles of fairness in ML for development (Awwad et al., 2020).

2. Stakeholder ecosystem design: the protocol intentionally maps diverse actors typically excluded from technical policy analysis (e.g., local government engineers, representatives from marginalized communities, Indigenous populations, technical domain experts, civil society advocates, and frontline implementers). This mapping necessitates structured stakeholder analysis methodologies to ensure systematic and justified inclusion, drawing on established typologies from fields like natural resource management (Reed et al., 2009). The creation of channels for marginalized voices (Fricker, 2007) and procedural justice ensure affected populations participate in validation.
3. Algorithmic recommendation as deliberative provocation: we position computational outputs not as prescriptions for action but as provocations for stakeholder deliberation (Fischer, 2012). Stakeholders scrutinize recommendations against local capacity, political economy, cultural acceptability, and historical equity concerns. Then, we formalize this qualitative, contextual intelligence as constraints or weights fed back into the model's reward function through Constrained Markov Decision Processes (Liu et al., 2021). This ensures subsequent iterations align with societal values determined through democratic processes. Fischer-Abaigar et al. (2024) document the critical need for this approach. ML models in public sectors often ignore distribution shifts, label bias, and historical policy biases. The risk is an automation of systemic discrimination when prioritizing accuracy over equity. In line with this, Fischer-Abaigar et al. (2024) advocate prioritizing decision quality over prediction through frameworks like counterfactual analysis and policy learning paired with stakeholder input. This is precisely BRIDGE-E's design.
4. Participatory policy simulations: reinforcement learning (RL) techniques facilitate scenario-based decision-making where policymakers and stakeholders engage in real-time adjustments and feasibility assessments before implementation. Here, we draw on frameworks for policy learning with constraints (Liu et al., 2021). These simulations allow stress-testing policies against diverse value priorities, revealing unintended consequences of optimizing narrow metrics. This follows Kauffman's (1993) principle that adaptive fitness is contextually determined by position within continuously evolving configurations.
5. Adaptive governance models: continuous refinement mechanisms incorporate real-world feedback and evolving socio-economic conditions. This enables genuine co-evolution between computational systems and governance processes rather than one-time tool deployment. This embodies the need for adaptive, con-

tinuous processes to address governance gaps in complex systems as demonstrated by Weitz et al. (2017).

**How This Addresses Structural Inertia and Builds Capacity**
The HITL protocol addresses Hannan and Freeman's (1984) structural inertia by creating mechanisms for institutional co-evolution rather than producing outputs for assimilation into existing routines. The system builds endogenous institutional capability by requiring stakeholder participation from inception. It also maintains hybrid development teams and creates feedback loops where policy outcomes refine computational parameters. This implements Valle-Cruz and García-Contreras's (2025) finding that AI-driven transformation in the public sector requires "an operational transformation, which includes human resources and know-how as the spearhead."

The structured deliberation respects complexity dynamics (Holland, 1995; Arthur, 1989, 1999). It acknowledges distributed knowledge across stakeholders and treats policy coherence as an emergent property of well-orchestrated interactions rather than a centrally computable optimum. Starke and Lünich (2020) identify this approach as necessary for democratic legitimacy. We need to maintain human decision-making authority through hybrid systems while leveraging computational intelligence for enhanced deliberation.

## 7.3 Integrated Architecture: Cross-Component Synergies

The three modules function as an integrated system where cross-component interactions create emergent capabilities addressing the governance-data-inequality nexus systemically:

- **Epistemic justice pipeline:** Data harmonization creates inclusive knowledge base → Coherence assessment detects conflicts respecting diverse ontologies → HITL validation ensures marginalized perspectives shape final recommendations.
- This pipeline implements Fricker's (2007) epistemic justice end-to-end.
- **Complexity-aware feedback loops:** HITL deliberations identify ontological gaps → Data harmonization expands knowledge graphs → Coherence assessment refines with augmented ontology → Recommendations improve.
- This creates genuine co-evolution following Kauffman's (1993) co-evolutionary dynamics and Levinthal's (1997) adaptation on rugged landscapes.
- **Energy justice by design:** Data harmonization ensures representation → Coherence assessment embeds equity constraints → HITL validation and contestability mechanisms → Recommendations optimize within justice boundaries.
- This acknowledges Jenkins et al.'s (2016) three justice dimensions as constitutive architectural features.
- **Capacity co-production:** Hybrid teams develop data harmonization protocols → Local practitioners interpret coherence assessments → Stakeholders lead HITL validation → Institutional ownership builds progressively.

- This addresses Hannan and Freeman's (1984) structural inertia through institutional transformation.

## 7.4 Technical Implementation Specifications

### *7.4.1 Computational Architecture*

**Data Harmonization Module**

- NLP: lightweight transformer architectures for policy document processing in resource-constrained settings.
- Knowledge graphs: entity extraction, relationship mapping, and ontology representation.
- Synthetic data: conditional generative models with community validation protocols.
- FAIR compliance: Structured metadata, interoperability standards.

**Coherence Assessment Engine**

- Semantic analysis: transformer-based models for cross-sectoral policy alignment.
- Network analysis: Graph Neural Networks (GNNs) for policy interaction modeling.
- XAI: model-agnostic explainability techniques for bias detection.
- Performance: optimized for low-resource computational environments.

**Recommendation System**

- HITL infrastructure: interactive dashboards, deliberation support tools.
- Simulation: Reinforcement Learning (RL) with Constrained Markov Decision Processes (Liu et al., 2021).
- Feedback integration: adaptive learning mechanisms, stakeholder input formalization.
- Governance: audit trails, contestability mechanisms, accountability structures.

### *7.4.2 Infrastructure Prerequisites in Resource-Constrained Environments*

Our analysis identifies critical prerequisites:

- **Computational requirements:** calibrated to documented resource constraints; specifications range from standard server capacity for basic semantic analysis to scalable cloud-based solutions for comprehensive cross-sectoral assessment. Lightweight AI models optimized for low-resource settings function efficiently in environments with limited computational power (Vuruma et al., 2024).
- **Network connectivity:** designed to acknowledge bandwidth limitations while supporting federated learning for data privacy via on-device training (Chalamala et al., 2022).

- **Data quality:** preprocessing must contend with the well-documented challenges of poor and scarce data prevalent in many developing countries. This creates foundational ethical risks for AI systems (Vibbi, 2024). Protocols therefore require efforts to structure policy documents in FAIR-compliant formats (Raza et al., 2024), potentially augmented by community-validated synthetic data generation to compensate for incomplete datasets.
- **Cloud deployment:** cloud-based options overcome local computational constraints while maintaining data sovereignty mandatory for national policy analysis.
- **Capacity building:** implementation requires training programs that build local ownership and context-specific capabilities. This approach aligns with principles of self-determined capacity development (Sokona, 2021).

### *7.4.3 Sustained Viability: Model Maintenance and Adaptation*

Long-term viability requires institutionalizing three adaptive maintenance processes:

1. **Optimized model retraining cycle:** maintaining model relevance as policies evolve requires strategic retraining decisions that balance computational costs against predictive performance degradation, aligning with cost-aware retraining frameworks (Mahadevan & Mathioudakis, 2024).
2. **Structured feedback integration protocol:** human-in-the-loop (HITL) validation and outcome monitoring generate feedback for model refinement. This continuous calibration aligns with constraint-based learning paradigms formalized through frameworks like Constrained Markov Decision Processes, surveyed in reinforcement learning literature (Liu et al., 2021).
3. **Adaptive security architecture:** the system must maintain modularity to incorporate AI advancements while implementing defenses against documented threats like training data poisoning, as demonstrated in machine learning (ML) security research (Cinà et al., 2023).

Full operational commitment therefore encompasses not only initial deployment but also permanent supervisory capacity for continuous adaptation.

## 7.5 A Complexity-Aware Approach to Validation and Research

### *7.5.1 From Metrics to Processes: Rethinking Validation*

Traditional validation seeks objective metrics against stable ground truths—an approach our theoretical framework reveals as epistemologically invalid for contested policy domains (Dunn, 2012). Policy coherence is an emergent property of complex adaptive systems (Holland, 1995, 1998; Kauffman, 1993), not a technical state we can easily measure. This aligns with Majone's (1989) seminal argument that policy-making is less a technical exercise of optimization and more a process of argumentation and persuasion within institutional fora, where the quality of deliberation shapes legitimacy and outcomes. Therefore, validating BRIDGE-E requires a fundamental shift. We need to move from asking *"Did the tool find the correct answer?"* to asking *"Did the tool foster more legitimate, adaptive, and equitable governance processes?"*

We propose a validation framework built on three principles derived from our complexity-ethics synthesis:

1. **Assess process legitimacy, not just output accuracy.** Following deliberative democratic theory (Gutmann & Thompson, 2004), evidence must come from qualitative assessment of deliberation within HITL protocols. This involves discourse analysis and participatory observation to evaluate epistemic inclusion and procedural justice. Tracking credibility differentials and hermeneutical resource distribution among participants is here essential (Fricker, 2007).
2. **Trace systemic adaptation, not static performance.** Consistent with theories of organizational learning (Levinthal & March, 1993) and path dependence (Arthur, 1989), validation requires longitudinal case studies. The focus is on documenting institutional co-evolution such as shifts in coordination patterns, routine adaptation, and endogenous capacity development over time.
3. **Conduct reflexive ethical interrogation, not compliance auditing.** Aligned with responsibility-by-design (Dignum, 2019), ethical assessment must be continuous and embedded. It requires epistemic auditing, systematically examining whose knowledge is represented or excluded. In addition, we need participatory evaluation of whether equity constraints align with stakeholder-defined justice claims (Jenkins et al., 2016).

This tripartite approach aligns with Williams' (2018) framework for legitimizing policy knowledge, which emphasizes "coherence" across process, "outcome," and the "identity" of the knowledge-producing system itself. Implementing this validation framework necessitates a mixed-methods approach (Burch & Heinrich, 2016), combining qualitative insights into deliberative processes with quantitative and longitudinal analysis of systemic outcomes.

### *7.5.2 Theoretical Boundaries as Generative Research Questions*

BRIDGE-E's implementation confronts constitutive boundaries. We frame these not as flaws but as generative research questions that define a responsible innovation agenda (Owen et al., 2013).

- **Boundary 1:** the interpretive nature of policy. Policy is a contested construct where ambiguity serves political functions (Stone, 2022).

The forum's task is to produce a formally agreed-upon set of new governance rules. These rules must establish:

- Formal contestability and appeals processes for algorithmic outputs, designed according to principles of reviewable automated decision-making (Cobbe et al., 2021).
- Epistemic inclusion protocols for data and deliberation.
- The authority structure of a standing oversight body.

### 7.6.2 *Phase 2: Collaborative Technical Adaptation and Stewardship Transfer*

The objective here is to conduct the technical adaptation of computational modules within the guardrails of the co-designed governance rules. We also need to ensure the irreversible transfer of stewardship to local institutions.

**Methodological Approach**

1. **Structured collaborative development:** we conduct technical work through structured, iterative collaboration, where external technical experts work in tandem with designated local counterparts in focused work cycles. We document all modifications in shared design logs (Simonsen & Robertson, 2012). This creates a transparent audit trail that links technical choices to governance principles.
2. **Governance-aligned validation:** each work cycle concludes with a review by a panel derived from the oversight body. This panel tests the adapted system's outputs against the agreed-upon governance rules, using locally salient policy dilemmas. This process enforces responsibility-by-design (Dignum, 2019) as a continuous, iterative practice.
3. **Stewardship portfolio development:** the local counterpart team builds a competency portfolio documenting their contributions to the design logs, their curation of local validation data, and their facilitation of community reviews. Here, we demonstrate proficiency through this portfolio and a final simulation where the team independently operates the system to analyze a novel policy conflict.

This two-phase protocol directly counters the pathologies theorized in Part II. Phase 1, through Ostromian analysis, addresses Arthur's (1989) path dependence by making "rules-in-use" explicit for deliberate redesign. Phase 2, through structured collaboration and stewardship transfer, counters Hannan and Freeman's (1984) structural inertia and Kauffman's (1993) co-evolutionary mismatches by building irreversible local capacity. The integrated deliberative and validation mechanisms throughout operationalize the necessary checks against Fricker's (2007) epistemic injustice.

## 7.7 Potential Impact and Cross-Sectoral Transferability

BRIDGE-E proposes a conceptual framework for policy coherence by integrating computational analytics with governance structures. We argue that we must continuously interpret algorithmic outputs within legitimate local governance processes. This directly addresses the implementation gaps in SDG 7 by designing systems that align with institutional realities, not just technical specifications (OECD, 2018).

The framework's modular, principle-based design suggests cross-sectoral transferability. Its logic provides a generalizable template for governance domains characterized by fragmentation and contested knowledge, such as public health (SDG 3) and education (SDG 4). Epistemic justice, structural inertia, justice, and complexity dynamics are indeed structural deficiencies in algorithmic systems.

This potential is grounded in middle-range theory (Merton, 1968), where abstract principles generate testable propositions across contexts. Realizing this potential requires the empirical validation outlined in Sect. 7.5. This begins with comparative pilots, model refinement for political dynamics, and the development of participatory adaptation protocols.

## 7.8 Conclusion: From Theory to Testable Proposition

We cannot determine policy coherence by computation alone. Human deliberation is essential for effective assessment. The BRIDGE-E framework proposes this fundamental shift. It structures computational methodologies not as arbiters of coherence, but as instruments to inform and support these deliberative processes.

This translates into a testable core proposition. The legitimacy and effectiveness of algorithmic policy support depend on its continuous integration into accountable, inclusive governance processes. We designed the validation protocol (Sect. 7.5) and the implementation methodology (Sect. 7.6) to interrogate this proposition, not to validate a technical product.

Consequently, BRIDGE-E reframes the challenge from technical optimization to institutional design for democratic sense-making. BRIDGE-E provides a template for governance to engage with complexity. If we want to govern algorithmic complex adaptive systems effectively, governance must itself become adaptive, distributed, and capable of learning. The final chapter examines this necessary evolution.

## References


Aagaard, T. S. (2015). Energy-environment policy alignments. *Washington Law Review, 90*, 1517. Available at: https://digitalcommons.law.uw.edu/wlr/vol90/iss4/2

Ananny, M., & Crawford, K. (2018). Seeing without knowing: Limitations of the transparency ideal and its application to algorithmic accountability. *New Media & Society, 20*(3), 973–989. https://doi.org/10.1177/1461444816676645

Arthur, W. B. (1989). Competing technologies, increasing returns, and lock-in by historical events. *The Economic Journal, 99*(394), 116–131. https://doi.org/10.2307/2234208

Arthur, W. B. (1999). Complexity and the economy. *Science, 284*, 107–109. https://doi.org/10.1126/science.284.5411.107

Awwad, Y., Fletcher, R., Frey, D., Gandhi, A., Najafian, M., & Teodorescu, M. (2020). *Exploring fairness in machine learning for international development*. CITE MIT D-Lab. https://d-lab.mit.edu/sites/default/files/inline-files/Exploring_fairness_in_machine_learning_for_international_development_04012020_pages.pdf

Bates, R. H. (1997). Area studies and the discipline: A useful controversy? *PS: Political Science & Politics, 30*(2), 166–169. https://doi.org/10.2307/420485

Bender, E. M., & Friedman, B. (2018). Data statements for natural language processing: Toward mitigating system bias and enabling better science. *Transactions of the Association for Computational Linguistics, 6*, 587–604. https://doi.org/10.1162/tacl_a_00041

Bennett, A., & Checkel, J. T. (Eds.). (2015). *Process tracing*. Cambridge University Press.

Burch, P., & Heinrich, C. (2016). *Mixed methods for policy research and program evaluation*. SAGE Publications, Inc. https://doi.org/10.4135/9781483398259

Chalamala, S. R., Kummari, N. K., Singh, A. K., et al. (2022). Federated learning to comply with data protection regulations. *CSIT, 10*, 47–60. https://doi.org/10.1007/s40012-022-00351-0

Cinà, A. E., Grosse, K., Demontis, A., Vascon, S., Zellinger, W., Moser, B. A., Oprea, A., Biggio, B., Pelillo, M., & Roli, F. (2023). Wild patterns reloaded: A survey of machine learning security against training data poisoning. *ACM Computing Surveys, 55*(13s), Article 294. https://doi.org/10.1145/3585385

Cobbe, J., Lee, M. S. A., & Singh, J. (2021). Reviewable automated decision-making: A framework for accountable algorithmic systems. In *Proceedings of the 2021 ACM conference on fairness, accountability, and transparency (FAccT'21)* (pp. 598–609). Association for Computing Machinery. https://doi.org/10.1145/3442188.3445921

Costanza-Chock, S. (2020). *Design justice: Community-led practices to build the worlds we need*. The MIT Press. https://doi.org/10.7551/mitpress/12255.001.0001

Debnath, K. B., & Mourshed, M. (2018). Challenges and gaps for energy planning models in the developing-world context. *Nature Energy, 3*, 172–184. https://doi.org/10.1038/s41560-018-0095-2

Debnath, R., Tkachenko, N., & Bhattacharyya, M. (2025). Enabling people-centric climate action using human-in-the-loop artificial intelligence: A review. *Current Opinion in Behavioral Sciences, 61*, 101482. https://doi.org/10.1016/j.cobeha.2025.101482

Deslandes Whitney, C., & Norman, J. (2024). Real risks of fake data: Synthetic data, diversity-washing and consent circumvention. In *Proceedings of the 2024 ACM conference on fairness, accountability, and transparency (FAccT'21)* (pp. 1733–1744). Association for Computing Machinery. https://doi.org/10.1145/3630106.3659002

Dignum, V. (2019). *Responsible artificial intelligence: How to develop and use AI in a responsible way*. Springer. https://doi.org/10.1007/978-3-030-30371-6

Dunn, W. N. (2012). *Public policy analysis: An integrated approach* (5th ed.). Routledge. https://doi.org/10.4324/9781315663012

Farr, M. (2018). Power dynamics and collaborative mechanisms in co-production and co-design processes. *Critical Social Policy, 38*(4), 623–644. https://doi.org/10.1177/0261018317747444

Farrell, H. (2025). AI as governance. *Annual Review of Political Science, 28*(1), 375–392. https://doi.org/10.1146/annurev-polisci-040723-013245

Firebanks-Quevedo, D., Planas, J., Buckingham, K., Taylor, C., Silva, D., Naydenova, G., & Zamora-Cristales, R. (2022). Using machine learning to identify incentives in forestry policy: Toward a new paradigm in policy analysis. *Forest Policy and Economics, 134*, 102624. https://doi.org/10.1016/j.forpol.2021.102624

Fischer, F. (2012). Participatory governance: From theory to practice. In D. Levi-Faur (Ed.), *The Oxford handbook of governance*. Oxford Academic., online edn, 18 Sept. 2012. https://doi.org/10.1093/oxfordhb/9780199560530.013.0032

Fischer-Abaigar, U., Kern, C., Barda, N., & Kreuter, F. (2024). Bridging the gap: Toward an expanded toolkit for AI-driven decision-making in the public sector. *Government Information Quarterly, 41*(4), 101976. https://doi.org/10.1016/j.giq.2024.101976

Floridi, L. (2013). *The ethics of information*. Oxford University Press. https://doi.org/10.1093/acprof:oso/9780199641321.001.0001

Floridi, L., & Taddeo, M. (2016). What is data ethics? *Philosophical Transactions of the Royal Society A, 374*, 20160360. https://doi.org/10.1098/rsta.2016.0360

Fricker, M. (2007). *Epistemic injustice: Power and the ethics of knowing*. Oxford University Press. https://doi.org/10.1093/acprof:oso/9780198237907.001.0001

Garud, R., Kumaraswamy, A., & Karnøe, P. (2010). Path dependence or path creation? *Journal of Management Studies, 47*(4), 760–774. https://doi.org/10.1111/j.1467-6486.2009.00914.x

Gómez-Baggethun, E., Corbera, E., & Reyes-García, V. (2013). Traditional ecological knowledge and global environmental change: Research findings and policy implications. *Ecology and Society, 18*(4), 72. https://doi.org/10.5751/ES-06288-180472

Gutmann, A., & Thompson, D. (2004). *Why deliberative democracy? STU-Student edition*. Princeton University Press. JSTOR, http://www.jstor.org/stable/j.ctt7t5w5

Häbel, S., & Hakala, E. (2021). Policy coherence for sustainable development and environmental security: A case study of European Union policies on renewable energy. *Environmental Policy and Governance, 31*(6), 633–646. https://doi.org/10.1002/eet.1962

Hannan, M. T., & Freeman, J. (1984). Structural inertia and organizational change. *American Sociological Review, 49*(2), 149–164. https://doi.org/10.2307/2095567

Hayek, F. A. (1945). The use of knowledge in society. *The American Economic Review, 35*(4), 519–530. http://www.jstor.org/stable/1809376

Holland, J. H. (1995). *Hidden order: How adaptation builds complexity*. Addison-Wesley.

Holland, J. H. (1998). *Emergence from chaos to order*. Oxford University Press. https://doi.org/10.1093/oso/9780198504092.001.0001

Jenkins, K., McCauley, D., Heffron, R., Stephan, H., & Rehner, R. (2016). Energy justice: A conceptual review. *Energy Research & Social Science, 11*, 174–182. https://doi.org/10.1016/j.erss.2015.10.004

Jobin, A., Ienca, M., & Vayena, E. (2019). The global landscape of AI ethics guidelines. *Nature Machine Intelligence, 1*, 389–399. https://doi.org/10.1038/s42256-019-0088-2

Katzenbach, C., & Ulbricht, L. (2019). Algorithmic governance. *Internet Policy Review, 8*(4). https://doi.org/10.14763/2019.4.1424

Kauffman, S. A. (1993). *The origins of order: Self-organization and selection in evolution*. Oxford Academic. https://doi.org/10.1093/oso/9780195079517.001.0001

Kindon, S., Pain, R., & Kesby, M. (2007). *Participatory action research approaches and methods. Connecting people, participation and place* (p. 260). Routledge.

Levinthal, D. A. (1997). Adaptation on rugged landscapes. *Management Science, 43*(7), 934–950. https://doi.org/10.1287/mnsc.43.7.934

Levinthal, D. A., & March, J. G. (1993). The myopia of learning. *Strategic Management Journal, 14*, 95–112. https://doi.org/10.1002/smj.4250141009

Lindgren, S., & Dignum, V. (2023). Chapter 14: Beyond AI solutionism: Toward a multi-disciplinary approach to artificial intelligence in society. In *Handbook of critical studies of artificial intelligence* (pp. 179–193). Edward Elgar Publishing. https://doi.org/10.4337/9781803928562.00019

Liu, Y., Halev, A., & Liu, X. (2021, August). Policy learning with constraints in model-free reinforcement learning: A survey. In *The 30th international joint conference on artificial intelligence (IJCAI)*. https://par.nsf.gov/servlets/purl/10313521

Lukes, S. (2005). *Power: A radical view* (2nd ed.). Palgrave Macmillan.

Mahadevan, A., & Mathioudakis, M. (2024). Cost-aware retraining for machine learning. *Knowledge-Based Systems, 293*, 111610. https://doi.org/10.1016/j.knosys.2024.111610

Majone, G. (1989). *Evidence, argument, and persuasion in the policy process*. Yale University Press.

Mangla, S. K., Srivastava, P. R., Eachempati, P., & Tiwari, A. K. (2024). Exploring the impact of key performance factors on energy markets: From energy risk management perspectives. *Energy Economics, 131*, 107373. https://doi.org/10.1016/j.eneco.2024.107373

Merton, R. K. (1968). *Social theory and social structure (Enlarged ed.)*. The Free Press.

Mittelstadt, B. (2019). Principles alone cannot guarantee ethical AI. *Nature Machine Intelligence, 1*, 501–507. https://doi.org/10.1038/s42256-019-0114-4

Moore, G. J., Bardagot, O., & Banerji, N. (2022). Deep transfer learning: A fast and accurate tool to predict the energy levels of donor molecules for organic photovoltaics. *Advanced Theory and Simulations, 5*, 2100511. https://doi.org/10.1002/adts.202100511

Nepal, R., Sharma, B., & al Irsyad, M. I. (2020). Scarce data and energy research: Estimating regional energy consumption in complex economies. *Economic Analysis and Policy, 65*, 139–152. https://doi.org/10.1016/j.eap.2019.12.002

Nilashi, M., Keng Boon, O., Tan, G., Lin, B., & Abumalloh, R. (2023). Critical data challenges in measuring the performance of sustainable development goals: Solutions and the role of big-data analytics. *Harvard Data Science Review, 5*(3). https://doi.org/10.1162/99608f92.545db2cf

Novelli, C., Taddeo, M., & Floridi, L. (2024). Accountability in artificial intelligence: What it is and how it works. *AI & Society, 39*, 1871–1882. https://doi.org/10.1007/s00146-023-01635-y

Ostrom, E. (2011). Background on the institutional analysis and development framework. *Policy Studies Journal, 39*, 7–27. https://doi.org/10.1111/j.1541-0072.2010.00394.x

Owen, R., Stilgoe, J., Macnaghten, P., Gorman, M., Fisher, E., & Guston, D. (2013). A framework for responsible innovation. In *Responsible innovation: managing the responsible emergence of science and innovation in society* (pp. 27–50). https://doi.org/10.1002/9781118551424.ch2

Raza, S., Ghuge, S., Ding, C., Dolatabadi, E., & Pandya, D. (2024). FAIR enough: Develop and assess a FAIR-compliant dataset for large language model training? *Data Intelligence, 6*(2), 559–585. https://doi.org/10.1162/dint_a_00255

Reed, M. S., Graves, A., Dandy, N., Posthumus, H., Hubacek, K., Morris, J., Prell, C., Quinn, C. H., & Stringer, L. C. (2009). Who's in and why? A typology of stakeholder analysis methods for natural resource management. *Journal of Environmental Management, 90*(5), 1933–1949. https://doi.org/10.1016/j.jenvman.2009.01.001

Schneider, A., & Ingram, H. (1993). Social construction of target populations: Implications for politics and policy. *American Political Science Review, 87*(2), 334–347. https://doi.org/10.2307/2939044

Simonsen, J., & Robertson, T. (Eds.). (2012). *Routledge international handbook of participatory design* (1st ed.). Routledge. https://doi.org/10.4324/9780203108543

Sokona, Y. (2021). Building capacity for 'energy for development' in Africa: Four decades and counting. *Climate Policy, 22*(5), 671–679. https://doi.org/10.1080/14693062.2020.1870915

Starke, C., & Lünich, M. (2020). Artificial intelligence for political decision-making in the European Union: Effects on citizens' perceptions of input, throughput, and output legitimacy. *Data & Policy, 2*, e16. https://doi.org/10.1017/dap.2020.19

Sterman, J. D. (2000). *Business dynamics: Systems thinking and modeling for a complex world*. McGraw-Hill.

Stone, D. A. (2022). *Policy paradox: The art of political decision making* (3rd ed.). W.W. Norton & Company.

Valle-Cruz, D., & García-Contreras, R. (2025). Towards AI-driven transformation and smart data management: Emerging technological change in the public sector value chain. *Public Policy and Administration, 40*(2), 254–275. https://doi.org/10.1177/09520767231188401

Vibbi, L. F. (2024). Poor data quality in sub-Saharan Africa and implications on ethical AI development. In S. Chiodo, D. Kaiser, J. Shah, & P. Volonté (Eds.), *Improving technology through ethics* (SpringerBriefs in applied sciences and technology). Springer. https://doi.org/10.1007/978-3-031-52962-7_7

Vuruma, S. K. R., Margetts, A., Su, J., Ahmed, F., & Srivastava, B. (2024). From cloud to edge: Rethinking generative AI for low-resource design challenges. *arXiv preprint arXiv:2402.12702*. https://doi.org/10.48550/arXiv.2402.12702

Weitz, N., Strambo, C., Kemp-Benedict, E., & Nilsson, M. (2017). Closing the governance gaps in the water-energy-food nexus: Insights from integrative governance. *Global Environmental Change, 45*, 165–173. https://doi.org/10.1016/j.gloenvcha.2017.06.006

Wieringa, M. (2020). What to account for when accounting for algorithms: A systematic literature review on algorithmic accountability. In *Proceedings of the 2020 conference on fairness, accountability, and transparency (FAT'20)* (pp. 1–18). Association for Computing Machinery. https://doi.org/10.1145/3351095.3372833

Wilkinson, M., Dumontier, M., Aalbersberg, I., et al. (2016). The FAIR guiding principles for scientific data management and stewardship. *Scientific Data, 3*, 160018. https://doi.org/10.1038/sdata.2016.18

Williams, K. (2018). Three strategies for attaining legitimacy in policy knowledge: Coherence in identity, process and outcome. *Public Administration, 96*, 53–69. https://doi.org/10.1111/padm.12385

Wilson, J. Q. (1989). *Bureaucracy: What government agencies do and why they do it*. Basic Books.

Yeung, K. (2018). Algorithmic regulation: A critical interrogation. *Regulation & Governance, 12*, 505–523. https://doi.org/10.1111/rego.12158

Yin, R. K. (2018). *Case study research and applications: Design and methods* (6th ed.). Sage.

Zhang, R., & El-Gohary, N. (2023). Transformer-based approach for automated context-aware IFC-regulation semantic information alignment. *Automation in Construction, 145*, 104540. https://doi.org/10.1016/j.autcon.2022.104540

Zhang, Y., Huang, T., & Bompard, E. F. (2018). Big data analytics in smart grids: A review. *Energy Informatics, 1*, 8. https://doi.org/10.1186/s42162-018-0007-5

Zheng, X., Wang, B., Zhao, Y., Mao, S., & Tang, Y. (2021). A knowledge graph method for hazardous chemical management: Ontology design and entity identification. *Neurocomputing, 430*, 104–111. https://doi.org/10.1016/j.neucom.2020.10.095

Zheng, Z., Zhou, Y.-C., Lu, X.-Z., & Lin, J.-R. (2022). Knowledge-informed semantic alignment and rule interpretation for automated compliance checking. *Automation in Construction, 142*, 104524. https://doi.org/10.1016/j.autcon.2022.104524

Zhu, W. (2025, July). AI-enabled policy pathways for green innovation in heavy polluters: An innovation ecosystem study with BERT, DRL, and GNN integration. In *Proceedings of the 2025 2nd international conference on big data and digital management* (pp. 282–289). https://doi.org/10.1145/3768801.3768848

# Chapter 8
# Governing Algorithmic Complexity

This concluding chapter synthesizes the empirical and theoretical findings presented throughout this book. We articulate their implications for rethinking governance in algorithmic societies. Building on the governance-data-inequality nexus diagnosed in Chap. 1 and the empirical patterns documented in Chaps. 4, 5 and 6, we advance a theoretical framework for governing algorithmic complexity through complexity-aware institutional design.

## 8.1 The Foundational Challenge: From Implementation Failure to Conceptual Mismatch

Consider a national energy ministry deploying a machine learning (ML) model to optimize grid investment. Trained on decades of consumption data from well-connected urban centers, the model systematically recommends against decentralized solar microgrids for underserved rural regions. Such a model computationally entrenches decades of spatial inequality. This is not a software error but a perfect execution of its design. Trained on biased data and shaped by institutional fragmentation, the model cannot overcome its intrinsic limitations. This vignette exemplifies the central paradox identified through our analysis. Computational tools designed to enhance coordination instead reveal fundamental governance inadequacies.

Traditional governance fails with algorithmic systems because conventional bureaucratic structures designed for reliability and accountability (Wilson, 1989; Hannan & Freeman, 1984) cannot accommodate systems exhibiting increasing returns, lock-in and co-evolutionary dynamics (Arthur, 1989, 1999; Kauffman, 1993; Holland, 1995, 1998). This incompatibility manifests through dependent data exclusions, institutional-technical temporal mismatches, and self-reinforcing fragmentation feedback loops.



R. Bouderhem, *A Computational Framework for Energy Policy Coherence*, SpringerBriefs in Energy, https://doi.org/10.1007/978-3-032-22978-6_8

First, algorithms inherit and computationally harden institutional fragmentation. Models trained on data from siloed agencies cannot comprehend cross-sectoral impacts. They generate recommendations that appear optimal within narrow domains. However, they perpetuate system-wide incoherence and institutional fragmentation (Chap. 4). Second, algorithms accelerate path dependence through computational lock-in. Early choices in data ontology and algorithmic design become progressively more difficult to reverse as institutional investments accumulate and technical systems become entrenched. Arthur (1989) identified this dynamic in technological systems, which is now operating at the scale of governance itself. Third, algorithms create co-evolutionary mismatches that outpace regulatory adaptation. Technical systems evolve along performance trajectories measured in months, while governance institutions evolve through political and bureaucratic processes measured in years or decades. This temporal disconnect, explained through Kauffman's (1993) framework of co-evolutionary dynamics, prevents the endogenous alignment necessary for effective regulation, manifesting what scholars identify as the "pacing problem" (Marchant, 2011).

These recurring failures are symptoms of a deeper conceptual rupture. Algorithmic systems operating within complex socio-technical environments systematically violate conventional governance premises. First, they produce emergent outcomes unpredictable from individual components (Holland, 1998). Second, they operate in constantly shifting landscapes far from equilibrium (Kauffman, 1993). They exhibit the self-organizing and co-evolutionary dynamics that complexity theory identifies as central to contemporary governance (Klijn, 2008). Third, these dynamics generate a pervasive architecture of adaptive, data-driven feedback. This architecture, exemplified by "hypernudges" (Yeung, 2017), displaces human deliberation. This enacts a form of testimonial injustice by creating a pervasive credibility deficit for non-algorithmic knowledge and hermeneutical injustice by making the logic of its influence opaque (Fricker, 2007). Finally, they embed ethical choices intrinsically within technical design parameters from data selection to optimization criteria (O'Neil, 2016).

The governance-data-inequality nexus is the material manifestation of this rupture. Fragmented governance produces fragmented data. This data, in turn, trains algorithms that reproduce and amplify the original fragmentation. The result is a self-justifying, computationally sophisticated feedback loop. The core challenge, therefore, is not to fit new tools into old institutional boxes, but to redesign governance to engage complexity on its own terms.

## 8.2 The Necessary Evolution: Governing Complexity with Complexity

To govern complex adaptive systems, governance must itself become a complex adaptive system (CAS). This represents not a desirable evolution but a logical implication of our diagnosis. If algorithmic systems exhibit CAS properties, effective governance requires commensurate capabilities. Drawing on the complexity-ethics lens developed throughout this work, we propose a three-dimensional evolution targeting the epistemic, institutional, and ethical foundations of governance. These dimensions are not merely additive but mutually constitutive. Complexity science provides the diagnostic framework for understanding governance failures, while ethical frameworks provide the normative guidance for redesign.

Governance must undergo an epistemic evolution from centralized expertise to distributed intelligence. Traditional governance relies on hierarchical expertise structures. In algorithmic systems, however, there is a distribution of relevant knowledge across technical specialists, domain experts, affected communities, and institutional actors. Governance therefore requires an infrastructure capable of integrating these diverse knowledge forms without imposing artificial hierarchies. Fricker's (2007) framework of epistemic justice becomes a functional requirement, not an ethical ideal. We must build institutional mechanisms that remediate testimonial injustice (the systematic credibility deficits assigned to marginalized knowers) and address hermeneutical injustice (the gaps in collective conceptual resources that render certain experiences inarticulable). Without this, governance remains blind to crucial dimensions of the systems it seeks to regulate. This is a fatal flaw in energy transitions where local and Indigenous knowledge is routinely excluded (Jenkins et al., 2016).

Concurrently, governance requires an institutional evolution from bureaucratic stability to adaptive capacity. Hannan and Freeman's (1984) theory of structural inertia explains why bureaucracies resist transformational change even when facing dysfunction. Reliability imperatives create self-reinforcing routines that filter out disruptive innovation. Governing algorithmic complexity demands institutional designs that actively balance what March (1991) identified as the core tension between exploration of new approaches and exploitation of existing capabilities. This tension is particularly acute in public sectors where accountability imperatives often dominate (Wilson, 1989). This necessitates adaptive architectures (e.g., regulatory sandboxes, experimental governance zones, and iterative policy cycles) where institutions can prototype, learn, and evolve their frameworks alongside the algorithmic systems they govern. This aligns with frameworks for agile decision-making (Room, 2011) and adaptive governance (Folke et al., 2005). It must also directly address the systemic characteristics of effective policy mixes: consistency, coherence, and credibility (Rogge & Reichardt, 2016; Kern et al., 2017).

Finally, governance demands an ethical evolution from external constraint to embedded reasoning. In algorithmic environments, ethical choices are baked into technical parameters. Floridi's (2013) information ethics and Dignum's (2019) responsibility-by-design paradigm thus require a shift from treating ethics as an external checklist to making it a constitutive, operational logic. As Lindgren and Dignum (2023) argue, moving beyond AI solutionism requires embedding ethical reasoning within the technical architecture itself. Governance must develop embedded ethical reasoning through continuous deliberative processes integrated with technical design cycles, accountability mechanisms distributed across human-machine networks, and justice metrics hard-coded into algorithmic optimization functions. Here, Jenkins et al.'s (2016) energy justice framework provides a vital template, moving justice from post-hoc evaluation to a constitutive design principle governing distribution, procedure, and recognition from the outset. This evolution moves beyond mere procedural legitimacy to achieve substantive outcome legitimacy, addressing both components of social justice (Tyler, 2000).

This triad represents a fundamental reconceptualization where complexity science and ethical frameworks mutually redesign governance. Complexity science reveals why governance fails (path dependence, structural inertia, co-evolutionary mismatches), while ethical frameworks prescribe how to redesign it (epistemic justice, adaptive capacity, embedded reasoning). The goal is no longer to impose stability and control, which is an impossible task that generates the documented pathologies, but to build capabilities for navigating complexity. As shown, we should be able to detect emergent patterns across scales (Scheffer et al., 2009), adapt through structured learning (Sterman, 2000), and maintain legitimacy through inclusive deliberation on contested values (Fischer, 2009). This is what we mean by governing complexity with complexity.

## 8.3 BRIDGE-E as Theoretical Exemplar: Instantiating Complexity-Aware Governance

BRIDGE-E demonstrates how we can computationally implement complexity-aware governance principles. Following design science research principles (Hevner et al., 2004), BRIDGE-E embodies the synthesis of complexity science and ethical theory, providing a concrete form to abstract principles. It shows that a complexity-aware and ethical governance is feasible and institutionally coherent by directly addressing the implementation challenges documented in Chap. 5.

Consider traditional data integration, which often imposes standardizing schemas that exclude non-conforming knowledge. This practice exemplifies Ostrom's (2007) "panacea problem" in institutional design and is a direct consequence of Arthur's (1989) path dependence in data infrastructure. BRIDGE-E directly counters this by creating an epistemic infrastructure based on knowledge graphs that maintain ontological diversity. This is a technical instantiation of polycentric governance (Ostrom, 2010), designed to break path dependence in data ontologies. By enabling communication across different knowledge systems without erasing their uniqueness, it actively expands hermeneutical resources (Fricker, 2007) and provides the informational foundation for intelligent adaptation. It also embodies Holland's (1995) principle that complex adaptive systems require diverse, interacting components to generate intelligent behavior. This approach directly counters the homogenizing tendencies documented in Chap. 4's analysis of current computational applications.

Where traditional policy assessment works in silos, reflecting the institutional fragmentation diagnosed throughout this work, BRIDGE-E implements systemic pattern recognition. Using graph neural networks, it models indirect, cross-sectoral policy interactions to reveal emergent conflicts and synergies. This operationalizes Kauffman's (1993) insight that systems evolve on "fitness landscapes" where components interact in complex ways, often producing unintended consequences from seemingly independent policies.

Additionally, it integrates the tripartite dimensions of Jenkins et al.'s (2016) energy justice as hard constraints within its algorithmic optimization. This ensures that distributional equity, procedural fairness, and recognition are not secondary considerations but primary rules that shape the solution space from inception. This directly implements Dignum's (2019) responsibility-by-design principle that ethical reasoning must be constitutive, not supplementary, and directly addressing the systematic ethical failures documented in Chap. 5.

Finally, BRIDGE-E confronts the structural inertia of hierarchical decision-making. Through human-in-the-loop (HITL) protocols, it creates a deliberative architecture that mandates stakeholder contestation and collective sense-making, drawing on theories of democratic expertise (Fischer, 2009). This embeds accountability and contestability directly into the system's structure, not as an external audit. The system's feedback mechanisms formalize stakeholder input to refine model parameters. This operationalizes systems thinking approaches to organizational learning (Sterman, 2000). It fosters genuine co-evolution between the computational system and the governance institution. This directly addresses the co-evolutionary mismatches identified in Chap. 6's institutional analysis.

BRIDGE-E makes tangible the synthesis at the heart of this work. Complexity science provides the diagnostic and design logic, ethical frameworks provide normative guardrails, and computational engineering provides implementation mechanisms. Computational systems can be the core infrastructure of a necessary governance evolution, moving beyond their role as tools for a failing paradigm. This evolution must directly address the empirical challenges documented throughout this book.

## 8.4 The Structural Imperative and the Governance Challenge

Our analysis reveals a constitutive mismatch with a logical consequence. If algorithmic systems exhibit the CAS properties of emergence, path dependence, and co-evolution (Holland, 1995. 1998; Arthur, 1989; Kauffman, 1993), and if we designed traditional governance architectures for linearity, hierarchy, and stability (Wilson, 1989), then their incompatibility is not accidental but structural. The governance-data-inequality nexus is the empirical signature of this mismatch. It reveals a computationally accelerated feedback loop that transforms institutional fragmentation into algorithmic bias and back again.

Therefore, the evolutionary triad outlined in Sect. 8.2 (epistemic, institutional, and ethical) is not a menu of optional enhancements. Governance must evolve as its environment evolves. The question shifts from whether governance should evolve to how it can cultivate these adaptive capacities against its own deeply ingrained inertial forces.

This presents the core governance challenge for algorithmic societies, which is to design reflexive institutions capable of learning and self-correction at a pace commensurate with the systems they steward. This challenge operates on two interdependent levels:

1. The internal challenge of metamorphosis: governance must overcome its own path dependence and structural inertia (Arthur, 1989; Hannan & Freeman, 1984) to develop the epistemic inclusivity, adaptive capacity, and embedded ethics required. This is an internal redesign problem, demanding new structures for deliberation, experimentation, and learning.
2. The external challenge of stewardship: simultaneously, these redesigned institutions must develop new regulatory practices to guide external algorithmic systems. These practices cannot be static rules but must be meta-regulatory. This will ensure we design algorithmic systems themselves for transparency, contestability, and alignment with values determined through democratic processes (Yeung, 2018; Dignum, 2019).

BRIDGE-E (Sect. 8.3) addresses both levels in an integrated form. It is both a proposal for an internal governance instrument (a tool for policy coherence) and an instantiation of the meta-regulatory principles (justice-by-design, deliberative HITL) that should guide external systems. It demonstrates that the internal and external challenges are two sides of the same coin. Governing algorithmic complexity requires governance to recompose itself around complexity-aware principles.

The task, then, is to move from this exemplar to a field of practice. This necessitates treating frameworks like BRIDGE-E not as solutions, but as testable hypotheses within the specific, contested contexts where the governance-data-inequality nexus persists.

## 8.5 Conclusion: From Exemplar to Praxis

This book has argued that the crisis of policy coherence (in energy and beyond) in the algorithmic age is a symptom of a deeper, structural crisis of governance. The traditional paradigm, designed for a world of simpler causal chains and slower feedback, is constitutively inadequate for systems characterized by emergence, path dependence, and co-evolution. The result is not inefficiency but a self-reinforcing cycle, the governance-data-inequality nexus, which computationally hardens fragmentation and injustice.

Consequently, the imperative is metamorphosis. Governance must evolve to become more adaptive, distributed, and ethically constitutive than conventional bureaucratic models. The synthesis of complexity science and ethical frameworks provides both the diagnostic lens to understand this necessity and the normative compass to guide the redesign. Complexity science explains the dynamics of failure; ethics prescribes the direction of evolution.

The BRIDGE-E framework shows that principles of epistemic justice, adaptive capacity, and justice-by-design can be computationally instantiated to support, not subvert, and legitimate democratic governance. Its value is as a testable exemplar and a methodological template. The rigorous validation protocol and implementation methodology established in Chap. 7 provide the means to validate its core proposition that computation must support deliberation rather than replace it.

The path forward is therefore neither prescriptive nor technocratic. It is experimental and iterative. It demands a shift in practice across domains:

- For policymakers and regulators, it means chartering "adaptation labs" to prototype new governance instruments, focusing on institutional learning and the development of meta-regulatory capacity.
- For engineers and designers, it mandates building systems where strengthening legitimate deliberation is a paramount design constraint, equal to predictive accuracy.
- For researchers, it necessitates the transdisciplinary, critical inquiry outlined in Chap. 7, treating governance innovation as a field of rigorous, embedded study.

The ultimate goal is not a perfectly coherent policy but a resilient governance system, one capable of sensing emergent incoherence, learning from failure, and deliberating under conditions of complexity and value conflict. If this work contributes to reorienting theory and practice toward this goal, by providing a diagnostic framework, a principled synthesis, and a testable exemplar, it will have achieved its purpose. The challenge of governing algorithmic complexity is the challenge of enabling governance's own necessary, and ongoing, evolution.

## References


Arthur, W. B. (1989). Competing technologies, increasing returns, and lock-in by historical events. *The Economic Journal, 99*(394), 116–131. https://doi.org/10.2307/2234208

Arthur, W. B. (1999). Complexity and the economy. *Science, 284*, 107–109. https://doi.org/10.1126/science.284.5411.107

Dignum, V. (2019). *Responsible artificial intelligence: How to develop and use AI in a responsible way*. Springer. https://doi.org/10.1007/978-3-030-30371-6

Fischer, F. (2009). Democracy and expertise: Reorienting policy inquiry. Oxford University Press. https://doi.org/10.1093/acprof:oso/9780199282838.001.0001, accessed 18 Dec. 2025.

Floridi, L. (2013). *The ethics of information*. Oxford University Press. https://doi.org/10.1093/acprof:oso/9780199641321.001.0001

Folke, C., Hahn, T., Olsson, P., & Norberg, J. (2005). Adaptive governance of social-ecological systems. *Annual Review of Environment and Resources, 30*(1), 441–473. https://doi.org/10.1146/annurev.energy.30.050504.144511

Fricker, M. (2007). *Epistemic injustice: Power and the ethics of knowing*. Oxford University Press. https://doi.org/10.1093/acprof:oso/9780198237907.001.0001

Hannan, M. T., & Freeman, J. (1984). Structural inertia and organizational change. *American Sociological Review, 49*(2), 149–164. https://doi.org/10.2307/2095567

Hevner, A. R., March, S. T., Park, J., & Ram, S. (2004). Design science in information systems research. *MIS Quarterly, 28*(1), 75–105. https://doi.org/10.2307/25148625

Holland, J. H. (1995). *Hidden order: How adaptation builds complexity*. Addison-Wesley.

Holland, J. H. (1998). *Emergence: From chaos to order*. Oxford University Press. https://doi.org/10.1093/oso/9780198504092.001.0001

Jenkins, K., McCauley, D., Heffron, R., Stephan, H., & Rehner, R. (2016). Energy justice: A conceptual review. *Energy Research & Social Science, 11*, 174–182. https://doi.org/10.1016/j.erss.2015.10.004

Kauffman, S. A. (1993). *The origins of order: Self-organization and selection in evolution*. Oxford Academic. https://doi.org/10.1093/oso/9780195079517.001.0001

Kern, F., Kivimaa, P., & Martiskainen, M. (2017). Policy packaging or policy patching? The development of complex energy efficiency policy mixes. *Energy Research & Social Science, 23*, 11–25. https://doi.org/10.1016/j.erss.2016.11.002

Klijn, E. H. (2008). Complexity theory and public administration: What's new? Key concepts in complexity theory compared to their counterparts in public administration research. *Public Management Review, 10*(3), 299–317. https://doi.org/10.1080/14719030802002675

Lindgren, S., & Dignum, V. (2023). Chapter 14: Beyond AI solutionism: Toward a multi-disciplinary approach to artificial intelligence in society. In *Handbook of critical studies of artificial intelligence* (pp. 179–193). Edward Elgar Publishing. https://doi.org/10.4337/9781803928562.00019

March, J. G. (1991). Exploration and exploitation in organizational learning. *Organization Science, 2*(1), 71–87. https://doi.org/10.1287/orsc.2.1.71

Marchant, G. E. (2011). Addressing the pacing problem. In G. Marchant, B. Allenby, & J. Herkert (Eds.), *The growing gap between emerging technologies and legal-ethical oversight* (The international library of ethics, law and technology) (Vol. 7). Springer. https://doi.org/10.1007/978-94-007-1356-7_13

O'Neil, C. (2016). *Weapons of math destruction: How big data increases inequality and threatens democracy*. Crown Publishing Group.

Ostrom, E. (2007). A diagnostic approach for going beyond panaceas. *Proceedings of the National Academy of Sciences, 104*(39), 15181–15187. https://doi.org/10.1073/pnas.0702288104

Ostrom, E. (2010). Beyond markets and states: Polycentric governance of complex economic systems. *American Economic Review, 100*(3), 641–672. https://doi.org/10.1257/aer.100.3.641

Rogge, K. S., & Reichardt, K. (2016). Policy mixes for sustainability transitions: An extended concept and framework for analysis. *Research Policy, 45*(8), 1620–1635. https://doi.org/10.1016/j.respol.2016.04.004

Room, G. (2011). *Complexity, institutions and public policy: Agile decision-making in a turbulent world*. Edward Elgar Publishing. https://doi.org/10.4337/9780857932648

Scheffer, M., Bascompte, J., Brock, W. A., Brovkin, V., Carpenter, S. R., Dakos, V., et al. (2009). Early-warning signals for critical transitions. *Nature, 461*(7260), 53–59. https://doi.org/10.1038/nature08227

Sterman, J. D. (2000). *Business dynamics: Systems thinking and modeling for a complex world*. McGraw-Hill.

Tyler, T. R. (2000). Social justice: Outcome and procedure. *International Journal of Psychology, 35*, 117–125. https://doi.org/10.1080/002075900399411

Wilson, J. Q. (1989). *Bureaucracy: What government agencies do and why they do it*. Basic Books.

Yeung, K. (2017). 'Hypernudge': Big data as a mode of regulation by design. *Information, Communication & Society, 20*(1), 118–136. https://doi.org/10.1080/1369118X.2016.1186713

Yeung, K. (2018). Algorithmic regulation: A critical interrogation. *Regulation & Governance, 12*, 505–523. https://doi.org/10.1111/rego.12158

# Index


**A**
Adaptive regulation, 20
Agent-based modeling (ABM), 20, 27, 51–52
AI ethics, 41
AI governance, 65, 70, 75, 79
Algorithmic bias, 7, 8, 12, 23, 42, 66–70, 113
Algorithmic governance, 90

**C**
Co-evolution, 19–21, 27, 52, 55–57, 66, 70, 82, 89, 90, 94–96, 99, 100, 113, 114
Complex adaptive systems (CAS), 7, 11, 19–20, 22, 23, 27, 28, 49, 51, 53–57, 61, 73, 74, 76, 77, 79, 83, 89, 90, 94, 98, 103, 110, 112, 113
Complexity economics, 7, 43, 90
Complexity-ethics framework, 36, 73–84
Complexity science, 7–9, 17–29, 43, 49, 56, 94, 111–113, 115
Computational policy analysis, 4, 7, 18, 26, 29, 33, 42
Computational social science, 9, 23

**D**
Data infrastructures, 33, 53–54, 57, 61–63, 66, 68–70, 74–79, 112
Developing countries, 4–6, 8–10, 12, 17, 24, 26, 35, 37, 39, 40, 42–44, 62, 75, 92, 98

**E**
Emergence, 7, 11, 19–21, 27, 28, 35, 36, 51, 56, 113, 114
Energy governance, 5, 6, 9, 17, 19–21, 37, 39, 40, 42, 43, 75, 81
Energy justice, 19, 21, 43, 45, 49, 50, 53, 56, 57, 61, 67, 71, 73, 76, 80, 82, 83, 89–91, 94, 96, 112, 113
Energy policy, 3–12, 23, 33, 35, 37, 38, 41–44, 49–58, 63, 70, 75
Epistemic injustices, 7–10, 20, 21, 23, 36, 43, 45, 49, 56, 61, 67, 78, 80, 83, 89, 101, 102
Epistemic justice, 19–21, 45, 63, 66, 73, 81, 83, 89, 91, 92, 96, 100, 103, 111, 112, 115

**G**
Governance, 3, 17, 33, 51, 61, 90, 109
Governance evolution, 113

**H**
Human-in-the-loop protocols (HITL), 113

**I**
Institutional analysis, 83, 101, 113
Institutional capacity, 3, 4, 6, 12, 17, 33, 63–66, 70, 78, 94
Institutional designs, 103, 109, 111, 112



R. Bouderhem, *A Computational Framework for Energy Policy Coherence*, SpringerBriefs in Energy, https://doi.org/10.1007/978-3-032-22978-6

**N**
Natural language processing (NLP), 4, 5, 9, 27, 37, 38, 40, 41, 43, 50–51, 53, 92, 93, 97
Network analysis, 27, 39, 52–53

**O**
OECD PCSD framework, 26, 28

**P**
Path dependence, 7, 10, 11, 19–22, 28, 36, 49, 52–54, 56, 57, 61–63, 65, 66, 68, 70, 71, 73, 74, 76, 77, 79–84, 89–93, 99–102, 110, 112–114
Policy coherence, 3–12, 17–19, 21–26, 28, 33, 37, 39–43, 55, 75, 89–103, 114
PRISMA-ScR, 34

**R**
Responsible AI, 8, 11, 19–21, 43, 45, 49, 67, 82, 91

**S**
Scoping reviews, 11, 29, 33–45, 49
SDG 7, 5, 6, 12, 17, 26, 27, 37–41, 43, 103
Structural inertia, 19, 20, 22, 28, 36, 49, 52, 54–57, 61, 63–66, 69–71, 73–84, 89, 93–97, 100, 102, 103, 111–114

**T**
Thematic saturation, 34, 36, 39, 41

The manufacturer's authorised representative in the EU is Springer Nature Customer Service Centre GmbH, Europaplatz 3, 69115 Heidelberg, Germany. If you have any concerns regarding our products, please contact ProductSafety@springernature.com

Printed and bound by CPI Group (UK) Ltd, Croydon, CR0 4YY
05/10/2026
02228034-0003